D0959928

BODY HEAT

BODY HEAT

TEMPERATURE AND LIFE ON EARTH

Mark S. Blumberg

HARVARD UNIVERSITY PRESS
Cambridge, Massachusetts, and London, England 2002

Printed in the United States of America

LIBRARY OF CONGRESS CATALOGING-IN-PUBLICATION DATA
Blumberg, Mark Samuel, 1961–
Body heat : temperature and life on earth / Mark S. Blumberg.
p. cm.
Includes bibliographical references and index.
ISBN 0–674–00762–X (alk. paper)
1. Body temperature–Regulation. 2. Animal heat. I. Title.
QP135 .B584 2002
571.7'6—dc21 2001051648

BOOK DESIGN BY JILL BREITBARTH

To Goldene and Herschel, for making it possible

To Jo, for making it worth it

And to Joseph K., Aja, and Millie, for just being you

CONTENTS

The cold has the philosophical value of reminding men that the universe does not love us. Cold as absolute as the black tomb rules space; sunshine is a local condition, and the moon hangs in the sky to illustrate that matter is usually inanimate.

—JOHN UPDIKE

Come on baby, light my fire.

—THE DOORS

INTRODUCTION

Temperature is an integral part of our daily lives, our rituals, and our language. Sitting at our kitchen counter on a winter morning, preparing to leave our warm house for the bitter cold and freezing rain outside, we rarely stop to consider how critically our evolution depended, and how critically our existence still depends, on the availability of a planet that is just the right distance from the Sun to support life. Instead, we gripe and grumble, refusing to acknowledge the obvious: Pluto is cold; Chicago in January is merely inconvenient.

Like a campfire on a winter night, the Sun is the primary source of warmth in our isolated region of the universe. As with a campfire, the heat emanating from the Sun decreases with distance, resulting in a continuous gradient of temperatures, moving outward like diminishing ripples on a pond, washing over the planets of our solar system. Thus Mercury, with an average distance from the Sun of 36 million miles, has a maximum surface temperature of approximately 400°C (752°F); while Pluto, with an average distance from the Sun of 3.7 billion miles, has a maximum surface temperature of approximately –225°C (–373°F). If planets were porridge, Goldilocks would have eaten neither.

Earth, however, is just right. Nestled between Venus and Mars some 90 million miles from the Sun, Earth presents a narrow thermal window of organic opportunity, a small opening between deep freeze and spontaneous combustion. But within this narrow window our world is overflowing with life. Animals have thrived in the temperate climes of North and South America, Europe, Africa, and Asia. Many species have also managed to make their living in the most intemperate terrestrial and aquatic environments. For example, on the dark and otherwise frigid floor of the Pacific Ocean, geologists discovered in 1977 the first of many hydrothermal vents where Earth's crustal plates have spread apart to allow hot, acidic gas to escape. Surrounding these thermal vents, more than one mile beneath the ocean surface, life has erupted at temperatures as high as 85°C (185°F). One denizen of these deep-water hot spots is a sulfur-consuming bacterium that forms the foundation of a food chain that comprises hundreds of exotic species, including blind shrimp, four-foot-long tube worms, and giant white crabs. Many other examples exist of life at the edge of organic possibility: fish with antifreeze in their blood inhabit Antarctic waters; desert rodents with super-efficient kidneys and respiratory passages thrive where leafy vegetation is the only source of water; and polar bears survive in the Arctic with the help of light-guiding shafts of hair and black skin for the maximal absorption of radiant energy. These are the true athletes of the extreme.

While some animals have evolved unique physiological capabilities to survive in extreme environments, our human ancestors colonized virtually every area of the globe from Africa to the Arctic by manipulating their surroundings to satisfy their physiological needs. For millennia we humans have clothed our naked skin with fur and leather and have built huts and homes to shield us from the elements. This ability to modify our external thermal conditions has been critically important for human expansion into the harshest terrestrial environments. When the Asian elephant migrated north to the Arctic, it evolved into a separate furred species—the woolly mammoth. Northward-migrating humans did not adopt the mammoth's strategy and grow fur, but rather compensated for their nakedness by hunting the woolly mammoth and other mammals for their coats. Today, of course, we shop.

Modifying our external environment to suit our thermal needs has been a continuous process. Each year we design, develop, and distribute commercial products that insulate us from the weather or help us to cope with its extremes: electric blankets, ceiling fans, down comforters, geothermal heat pumps, air conditioners, synthetic insulation for winter coats, toe warmers for ski boots. The demand for such products is unquestioned: each year Earth's human inhabitants spend billions of dollars on thermal products so that we can stay warm without shivering and stay cool without sweating.

Sometimes, as in 1980 when a heat wave hit the United States and 1,700 people died, or in July 2000, when a high-pressure area in southeastern Europe trapped hot air flowing from the Sahara desert, producing temperatures as high as 45°C (113°F) and resulting in the deaths of many people and livestock, we are not prepared either physiologically or behaviorally to cope with serious thermal challenges. Such events are particularly threatening to those who either cannot afford adequate thermal protection or cannot respond adequately to changes in the thermal environment—the poor, the very young, and the very old. For example, people eighty-five years of age and older are ten times more likely to die of hyperthermia than people fifty years of age or younger; the elderly are also three times more likely to die of hypothermia. (For now, let's define *hyperthermia* and *hypothermia* as unregulated increases and decreases in body temperature, respectively, such as when we exercise on a hot day or swim in frigid waters; as we will see, not all changes in body temperature fall into these categories.)

Creating and maintaining comfortable environments that protect us from extremes of hot and cold is a costly enterprise. For example, according to the U.S. Department of Energy, in 1997 the United States committed 55 percent of its total energy budget to home heating and air conditioning (an additional 19 percent was devoted to heating water). In total, 99 percent of American homes have some form of home heating system, and residents spend $42 billion each year on the energy to keep these systems running. The corresponding numbers for air conditioning are 73 percent and $10 billion, and in the hotter southern regions of the United States 93 percent of homes have air conditioning. The commitment of such a large proportion of a nation's financial and energy resources to the maintenance of the thermal environment testifies to its enormous importance.

Despite our intimate relations with temperature, our intuitions about it are remarkably poor, leaving us vulnerable to charlatans who seek to convince us that they have special powers. Many of us have seen mystics walking barefooted across hot coals—supposedly a demonstration of the mental discipline of these "fire walkers." Similarly, some Tibetan monks appear otherworldly in their resistance to the elements when they wrap themselves in cold, wet sheets without any apparent discomfort. We think we understand how hot coals and icy sheets should affect the human body, but we are wrong. In fact, any of us can walk across hot coals without pain if we don't dawdle, because wood, as a poor conductor of heat, requires a full second of contact with the foot to transfer sufficient heat to burn it (I'm still waiting to see someone walk over a bed of hot copper). Similarly, anyone—monk or not—can wrap himself in frigid sheets without becoming hypothermic because water is also a poor conductor of heat.

Temperature has played a central role in some of history's most significant events. For example, in June 1812, when Napoleon began his military campaign against Russia, he crossed the Polish-Russian border with about 422,000 men. They were dressed for summer. When the French army entered Moscow shortly after a costly battle at Borodino on September 7, the capital city had been largely destroyed by the retreating Russians. By this time Napoleon's army had dwindled to only 100,000 men. As the Russians withdrew from Moscow, an early and bitterly cold winter arrived that would prove devastating for the French. Over the succeeding weeks, temperatures plunged to –30°C (–22°F). Only 10,000 of Napoleon's soldiers survived to limp across the Russian border in November. The most powerful army of its time had been destroyed by poor insulation.

Temperature also played a central role in one of the great scientific debates of the nineteenth century. In 1859 Charles Darwin introduced his theory of natural selection to the world in his book *On the Origin of Species*. Darwin's work—inspired by the economic insights of Adam Smith and Thomas Malthus and grounded on a solid foundation of observation and logical argument—provided the first scientifically credible account of the evolution of life on Earth. Of course, not everyone was convinced. Among the many attacks on Darwin, one came from the eminent physicist William Thomson (who would become Lord Kelvin and later gain immortality by having the Kelvin temperature scale named after him). Thomson noted from his observations of the temperature of the Earth that, given the known laws of cooling objects, our planet was too young for Darwin to be correct.

Thomson had no qualms about declaring that his evidence undermined the likelihood of evolution because he also objected to the theory of evolution on theological grounds. Regardless of his personal beliefs, however, his logic was perfectly sound. Indeed, forensic scientists today use the same logic to establish the time of a person's death. The method involves little more than knowing the rate at which bodies cool and tracking back the current body temperature to the time when it would have been approximately 37°C (98.6°F). For an average body under normal conditions, body temperature decreases about 0.8°C (1.4°F) per hour; if one wants to be more accurate, one can take into account factors that govern the rate of heat loss from a body, such as body size, amount of clothing, and temperature of the air. But this method only works and (obviously) is only needed if the person is dead and has therefore ceased to produce heat.

Like a forensic scientist examining a corpse, Thomson set about to determine the age of the Earth. To do this, he measured the Earth's temperature at various depths below the surface to determine its rate of cooling (approximately 20–40°C, or 68–104°F, per kilometer). Then he calculated how long it would take for an object the size of the Earth to exhibit such cooling properties at its surface if it began in a molten state. On the basis of all the available evidence, Thomson concluded in a paper delivered to the Royal Society of Edinburgh in 1862 that the Earth was most likely 98 million years old. Thomson's pronouncement, made within three years of the publication of *The Origin,* had a devastating impact on Darwin and his theory: first, Darwin had presumed that many more millions of years would be required for the slow evolutionary process from single-celled organisms to humans; second, despite his fame as a naturalist, Darwin was in no position to dispute Thomson's findings on a matter of physics. Therefore Darwin was deeply troubled by the "odious spectre" of Thomson, and he attempted to address Thomson's criticism in subsequent editions of *The Origin*.

The method used by Thomson works well when applied to dead objects. Unfortunately for Thomson, reports of the Earth's death were greatly exaggerated. As we learned upon the discovery of radiation by Henri Becquerel and the Curies in the last few years of the nineteenth century, radioactive minerals like uranium produce heat, which means that our planet differs from a corpse. The fact that the Earth produces as well as loses heat significantly altered Thomson's calculations (the Earth is now estimated to be 4.6 billion, not 100 million years old) and restored the possibility of an ancient Earth demanded by Darwin's theory. Unfortunately for Darwin, the resolution of the controversy came long after his death in 1882.

Around the same time that Thomson was disturbing Darwin's sleep, a French physiologist named Claude Bernard was describing the *milieu intérieur*, or internal environment, of animals. Just as the Earth produces heat that slows its rate of cooling, so do we. But we do more: as Bernard pointed out, we maintain our temperature within a narrow range, and as we now know, we do so using a wide variety of mechanisms. We can shiver when we are cold and sweat when we are hot; we can dilate and constrict blood vessels in our skin to increase and decrease heat loss; we can curl up in a ball or sprawl, spread-eagled, on a cold floor; we can pull up the covers, drink hot fluids, put on warm clothes, jump in a pool, turn the dial on the thermostat, consult the Weather Channel, and do a variety of other things to cope with changes in our thermal environment. Some of these mechanisms are physiological (such as shivering) and some are behavioral (such as putting on clothes), but they are all weapons in our arsenal for fighting the thermal wars in which we are engaged each day.

As decades of scientific investigation have made clear, our body temperature has a fundamental influence on nearly all aspects of biological function, including the expression of genes, the activity of enzymes, the rate and force of contraction of muscles, the firing of neurons, and the habitats in which we can live. Sometimes those habitats change dramatically, as when a large asteroid apparently crashed into our planet 65 million years ago, creating a blast of heat followed by a dust cloud that blocked out sunlight for many months, leading to plummeting temperatures and drastic reductions of photosynthesis by green plants. The result was the extinction of the dinosaurs (not to mention two-thirds of the planet's marine and terrestrial species), which opened the diurnal (daytime) niche to a group of small, rodent-like, and previously nocturnal mammals—our ancestors. These ancestors survived in part because they were able to regulate their body temperature, even in the cold that followed the asteroid's devastating impact.

As those ancestors evolved and developed spoken and then written communication, thermal experiences also shaped and modified their language. Today, in English, we use thermal words to add spice or irony to our descriptions of objects or events, as in "hot topics" and "cold war," and we use thermal metaphors to describe the behavior of others, as in "he has a hot head," "she's warm-hearted," "he's boiling with anger," "he got cold feet," and "he's a cold-blooded killer." In the realm of sexual attraction, thermal words play a conspicuous role, as in "I've got the hots for you" or "he's cold." Consistently, when we use such expressions, warm is judged good and cold is judged bad. As we will see in Chapter 5, there is reason to believe that these judgments may derive from our infantile dependence on warmth provided by our mothers.

In the twenty-first century as in the age of the dinosaurs, temperature shapes life on Earth in myriad ways. Every animal has problems related to temperature: to find or avoid heat; to produce it when there is too little and dissipate it when there is too much; to direct the heat where it is needed; and to regulate it as efficiently as possible. As we will see, the solutions to these problems utilize some of the most elegant and creative mechanisms that the evolutionary process has devised. From the ability of a gazelle to outrun a hungry lion by cooling its brain to the ability of wading birds to stand with their feet in cold seawater for hours—literally from head to toe—we animals commit sizable portions of our time and energy to satisfying our thermal needs. Indeed, thermoregulation is serious business. When we are too hot or too cold, we are able to focus on little else but our lack of thermal comfort. But when we have achieved such comfort, we are free to do all the other things that make life worth living, such as reading a book.

1 TEMPERATURE: A USER'S GUIDE

THE LAWS OF PHYSICS form the foundation for all of science. Just as U.S. state laws cannot violate U.S. federal laws, the laws of chemistry and biology cannot violate the laws of physics. Animals, including most humans, may be unaware of the biological, chemical, and physical laws that govern our world, but the evolutionary process is not. Therefore, when they fly, animals obey the laws of gravity and aerodynamics. Similarly, when they thermoregulate, animals obey the laws of thermodynamics.

THERMODYNAMICS AND THE CONCEPT OF TEMPERATURE

As recently as two hundred years ago, physicists thought there was a substance called *caloric* that provided objects with their heat. Caloric, it was believed, flowed through an object like water between rocks in a stream. These scientists also thought that the amount of caloric in the universe never changed—in other words, that the quantity of heat in the universe was *conserved.* Our understanding of these processes underwent a major transformation when we came to view heat and work as two methods of transferring energy, and to believe that it is energy that is conserved, not heat. The proposition that there is a constant, unchanging quantity of energy in the universe is now known as the First Law of Thermodynamics.

Most of us are familiar, at least implicitly, with the notion that heat and work are somewhat interchangeable. For example, steam engines use heated water to produce work. Conversely, work can be converted into heat through friction, as when we vigorously rub two sticks together to make a fire. But there are limitations on their interchangeability. Imagine a river flowing downhill toward a waterwheel. Because all of the molecules in the river's water are flowing coherently (that is, in the same direction), they are able to turn the waterwheel and, by doing so, to perform work. In contrast, imagine what would happen if the waterwheel were placed in a lake outfitted with water jets that created a random churning of the water; in this case, the water molecules would not be moving coherently and therefore the waterwheel would not move and work would not be done. Instead, the churning of the water would generate only heat.

Thus the coherent energy present in a flowing river and the incoherent energy present in a churning lake illustrate the connection between order and work and the connection between disorder and heat.

A steam engine—the mechanical device *par excellence* not only of the industrial revolution but also of the physicists who studied thermodynamics in the nineteenth century—uses heat to produce steam that is in turn organized coherently in a pipe (as water is channeled in a river) to drive a turbine. Intensive investigation of the steam engine led to a depressing insight, now known as the Second Law of Thermodynamics, that has forever quashed any hope of developing a perpetual motion machine. The Second Law has been stated in a number of ways, but its main points can be summarized as follows: First, while it is true that heat can be channeled to produce work (as in a steam engine), this conversion of heat to work inevitably entails some waste, and this waste cannot be fully overcome through better engineering. Second, in the absence of other influences, heat flows downhill as water does; just as water does not spontaneously defy gravity and flow uphill, cold objects do not spontaneously heat up. To defy either gravity or the Second Law requires expending energy.

The Second Law of Thermodynamics is perhaps best known for the statement that the *entropy*, or disorder, of the universe increases steadily over time. Putting this idea together with the

First Law yields the following picture: The universe began and continues with a constant amount of energy, but the quality of this energy changes over time. Like river water flowing downhill past a waterwheel, turning the wheel and thereby producing work, the quality of energy in the universe continually declines from high-quality, work-producing energy to mere heat. In the end, although we may fight the slide toward disorder at the local level—this is what power plants and animals do—the universal trajectory toward disorder is unavoidable.

We are now ready to better understand the nature of heat and, in turn, what we mean by temperature. First, consider the process by which sugar makes its way from the sugarcane plant to your table. Initially the sugar is in a disordered state, dispersed throughout all the plants in the field. With manual labor and refining, the product, now in the form of sugar cubes, is packaged and transported to a grocery store, where you buy it. Ultimately, on a cold winter night, you prepare a cup of hot tea. You drop a cube into the cup of tea and, with some stirring, the sugar diffuses throughout the liquid until all the sugar molecules are evenly distributed. The highly ordered arrangement of sugar molecules in the cube (a low-entropy state) has given way to a disordered arrangement of sugar molecules in the tea (a high-entropy state). Every sip is equally sweet.

Cups of tea can be both sweet and hot, and these two qualities share an interesting property. Sweetness reflects the amount of sugar within a specific volume of fluid; a sugar cube dissolved in a cup of tea will result in sweeter tea than a sugar cube dissolved in a gallon of tea. In other words, sweetness is related to the *concentration* of sugar in a fluid. The same is true for temperature. For example, let's place a sugar-cube-sized piece of copper into an oven at, say, 500°C, and give it some time to heat up. Next, let's place this copper cube into a cup filled with room-temperature water. Like the sweetness from the sugar cube, the heat from the copper cube is eventually distributed evenly throughout the liquid (we are assuming here that the cup is well insulated so that the heat inside it is not escaping rapidly to the external environment). Moreover, as with the sugar cube, a heated copper cube dropped into a cup of water will result in hotter water than the same cube dropped into a gallon of water. Although the same *quantity* of heat is placed in each container, the resulting temperatures differ because temperature is a measure of the *concentration* of heat. To put it another way, temperature measures the local concentration of incoherently moving molecules in a substance.

The development of methods to measure the concentration of heat—the temperature—of objects required the contributions of numerous scientists. Although Galen, the second-century anatomist and physiologist, contrived a theory of human health based upon the idea that individuals differ in their relative proportions of hot and cold, he possessed no reliable method of measuring these proportions.

Galileo is credited with inventing the thermometer in the late sixteenth century, but his device, which took advantage of the well-known observation that air expands when heated, was really only useful for measuring the presence of heat. In 1654 Ferdinando II, Grand Duke of Tuscany and patron of Galileo, improved upon Galileo's open-air design with a sealed-in-glass, alcohol-filled instrument that included a demarcated scale; these innovations earned him the credit for the invention of the modern thermometer. Other models were quickly devised, many of them utilizing the extraordinary glass-blowing skills of the artisans of northern Italy. Despite this flurry of technical innovation, however, no one at this time merged the two features that would become the hallmarks of a true thermometer—a reliable instrument and a useful, reproducible scale.

Daniel Fahrenheit, a Polish-born scientist working in the Netherlands in the early eighteenth century, was the first to combine those two necessary features: a reliable mercury-in-glass thermometer (Fahrenheit switched from alcohol to mercury because the variability in alcohol's purity made it difficult to obtain reproducible results, a problem that did not arise with mercury); and a reproducible scale. For his scale, Fahrenheit, like others before him, chose to calibrate his instrument using two arbitrary fixed points. Around 1720 Fahrenheit established his scale's base value as the freezing point of water, which he designated as 30°F, and established the upper value as human body temperature, which he initially designated as 96°F. In time, he settled on values of 32°F and 212°F for the freezing and boiling points of water, and divided the difference into 180 equal parts. Subsequent scales differed only in the choice of the two fixed points (see Table 1). The centigrade scale, published by Anders Celsius in 1742, uses the freezing point of water (0°C) and the boiling point of water (100°C) with 100 equal degree divisions between these two extremes. The Kelvin scale uses the degree divisions –of the Celsius scale but places the zero point at –273.2°C (–459.7°F). The advantage of the Kelvin scale is that its bottom fixed point is not arbitrary as in all other scales but is, rather, tied to the lowest possible temperature in the universe, a value that corresponds to the cessation of molecular motion. Therefore the Kelvin scale contains no negative values, and for this reason it is referred to as an absolute scale.

To perform the precise and highly localized thermal measurements needed for scientific experiments, mercury thermometers cannot be used because they are too slow, bulky, and difficult to read. Therefore physiologists have employed a number of ingenious methods for the measurement of temperature. For example, thermocouples are constructed of two wires of different elements or alloys that are soldered together at a junction; the temperature of this junction varies in a systematic way with

Table 1 The three major temperature scales

Condition	Fahrenheit (°F)	Celsius (°C)	Kelvin (°K)
Molecular motion ceases	–459.8°	–273.2°	0.0°
Water freezes	32.0°	0.0°	273.2°
Typical room temperature	72.0°	22.2°	295.4°
Mean human body temperature	98.6°	37.0°	310.2°
Water boils	212.0°	100.0°	373.2°

Temperature conversions: °F = (°C × 9/5) + 32; °C = (°F – 32) × 5/9; °K = °C + 273.2

the electrical resistance measured across the junction. Thermocouples are favored by physiologists because they are sensitive, accurate, and flexible, thus allowing them to be used for the measurement of a variety of physiological temperatures, including heart and brain temperatures.

THE PHYSICS AND PHYSIOLOGY OF HEAT TRANSFER

The physics of heat transfer is straightforward, involving only three primary processes, *conduction, radiation,* and *evaporation,* and one auxiliary process, *convection.* As we review these processes, it may be useful to refer to Figure 1, which illustrates, for a horse wandering in a field on a sunny day, the

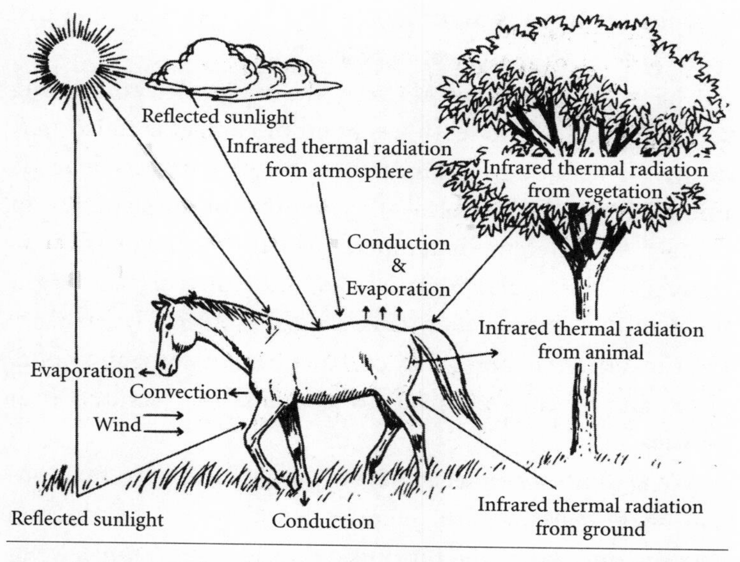

Figure 1 The myriad pathways by which heat is exchanged between an animal and its environment.

myriad ways in which heat is exchanged between an animal and its environment.

CONDUCTION

When one object (A) comes into contact with a second object (B), the two objects will eventually come to have the same temperature, a state known as thermal equilibrium. If, when they first come into contact, object A is hotter than object B, then object A will lose heat to object B.

The transfer of heat between the molecules composing solids, liquids, and gases at the point of contact is called *conduction.* The rate at which heat flows between objects depends upon just a few variables—the amount of surface area in contact, the difference in temperature between the two objects (that is, the *thermal differential*), and the ease with which heat flows in the material that makes up the objects (metals are poor insulators because they conduct heat easily, while wood, water, and fat are good insulators because they conduct heat poorly).

We and other animals, often without realizing it, manipulate these variables to increase or decrease conduction between our bodies and the outside world. For example, when walking on a hot surface, we may walk on tiptoe to minimize the area of contact and thereby decrease the conduction of heat to our body. Similarly, a lizard darting along the sun-heated stone wall of a Caribbean villa will typically raise its head, torso, and tail to accomplish the same end. When we wear oven mitts, we are increasing our insulation (and thereby decreasing the conduction of heat) at the point of contact with the hot pan. All of these are behavioral adjustments.

Much of an animal's heat exchange occurs at the interface between skin and air, and quite a bit of evolutionary creativity has been devoted to controlling this exchange. With regard to controlling conductive heat loss, there are two main solutions: blubber (below the skin) and fur or feathers (above the skin). Blubber, which is basically a deposit of fat beneath the skin, is used by marine mammals such as whales and seals to decrease the transfer of heat from their bodies to the frigid waters in which they swim. In fact, the harp seal is so effective at insulating itself against the cold with blubber that, even in icy Arctic waters, it can often maintain its body temperature without increasing its heat production.

External skin coverings such as fur or feathers are also used to control conductive heat loss. Mammals and birds independently evolved these solutions, as is clear from the very different compositions of a mink's fur and a duck's feathers. Although our intuition tells us that it is the covering itself that provides the insulative benefits, in fact it is the layer of air trapped by the fur and feathers that makes these coverings effective. Indeed, animals adjust the volume of this layer of air by flattening or erecting their fur or feathers depending on the needs of the moment. A chilled cat, for example, will erect its fur and thereby trap a layer of air above the surface of its skin. In contrast, when a cat is overheated, it will flatten its fur against its body to minimize the size of this insulative layer of air. Notice that the amount of fur has not changed, but rather the ability of the fur to trap air.

Humans have been called the naked ape for obvious reasons. We are, however, not completely naked: in addition to the patches of hair on our head and around our genitals, we have retained variable amounts of hair on many other parts of our body, including our arms, legs, and chest. We have all experienced the sensation of goose bumps—each bump is the site of a hair follicle—when we are cold. Why do we get goose bumps? They certainly can't have the function of retaining heat, because even the hairiest among us do not have enough hair to influence heat loss. In fact, goose bumps serve no function today, but they do provide modern evidence of our ancient condition of being, like all other apes, hairy. Goose bumps, in other words, are a vestigial character, an evolutionary holdover from a time when our fur coats were grown, not purchased. Today we compensate for our nakedness by wearing clothes and, as those of us living in cold climates have learned, by wearing layers of clothes as a way to trap layers of air.

(Goose bumps also occur in another important context. When heroin addicts abruptly quit using the drug, they experience a variety of withdrawal responses, including nausea, anxiety, fever, increased blood pressure, and goose bumps. This last symptom helped to inspire the phrase "quitting cold turkey.")

The insulative properties of fur and feathers are diminished greatly in water, because water disrupts the layer of air that is essential to retard heat loss. For animals that can avoid getting wet during the winter, this is not a problem. But for polar bears it is a serious one, because their surf-and-turf way of life in the Arctic requires that they swim in icy waters and travel over frozen land for food. Their solution? A thick coat of fur *plus* a layer of blubber. (Polar bears possess other adaptations as well—more about them later.)

CONVECTION

The transfer of heat by conduction is aided by convection, an auxiliary process that entails the transmission of heat from one location to another through its movement in a gas (for example, air) or a liquid. Perhaps the most familiar example is a fan blowing air over our skin. As we are all aware, fans don't cool air (if they did, we wouldn't need air conditioners) but rather provide us with the sensation of cool air. The sensation arises because the fan blows the layer of air, warmed by our skin and now insulating us against further loss of heat, away from us so that a cooler layer of air can take its place and accept the next conductive transfer of heat from our body. It is the acceleration of conductive heat transfer by convection that we sense as the cooling effect of the fan.

Sometimes convection can work against us. Consider the life-threatening situation posed when a ship sinks in frigid water. For the purposes of avoiding hypothermia, is it best to swim or tread water? As it turns out, you will stay warmer by treading water with your arms and legs as close to your body as possible, because swimming increases the convective movement of water over the body surface and therefore accelerates the conductive transfer of heat from the skin surface to the cold water.

Animals use convection in a number of creative ways. African elephants flap their large, thin ears when they are hot, increasing the convective movement of air over the skin of the ears and thereby cooling the blood coursing through these surfaces, which are highly vascularized—that is, plentifully supplied with blood vessels. One of the most essential uses of convection was made possible by the recruitment of the circulatory system as a thermoregulatory device. When we humans are hot, we dilate the blood vessels in the surface of our skin to allow the warm blood from the core of our body to flow to the surface and give up its heat to the outside air; as a result, we flush. In contrast, when we are cold, we constrict the blood vessels in the surface of our skin to conserve our body heat; as a result, we look pallid.

This simple convective mechanism is vital for many animals, whether living in hospitable or harsh environments. Recall the harp seal with the thick layer of blubber under its skin. What does a seal do when it has left the frigid water and is lying under the hot summer sun? How does it avoid becoming dangerously overheated without having to reenter the water to cool down? A seal can't simply push the blubber aside to let its body heat escape. So instead it dilates blood vessels that pass through its layer of blubber on the way to the skin, thus increasing the transportation of heat from the body core to the surface, where it is lost to the cool outside air.

RADIATION

Conduction entails the transfer of heat through direct contact between the atoms within a physical body. But not all heat transfer requires contact. You need not actually touch the iron skillet on your stove to know that it's hot—placing your hand a few inches away from it will do. The heat emitted by the skillet is a form of electromagnetic radiation (or light) that is normally beyond the sensitivity range of our visual system in the infrared part of the spectrum. At higher temperatures, however, the wavelength of the light decreases, and there comes a point when it is no longer invisible to the naked eye. Thus if your stove were able to heat the skillet to 1,300°C (2,372°F) it would glow red, and with further heating it would glow yellow and then white.

Although humans are blind to infrared radiation, other animals are not. For example, many snakes, including pythons, pit vipers, and rattlesnakes, possess dimples between the eyes and nostrils that contain receptors sensitive to infrared radiation. Snakes use the thermal information provided by these *pit organs* to detect prey and to organize successful predatory strikes.

As with conduction, heat loss through radiation occurs at surfaces when there is a difference between the internal body temperature and the external environmental temperature. Therefore animals can control the loss of radiant heat from their bodies through changes in posture, such as when a cat curls up in a tight ball and tucks in its paws and tail, that alter the amount of exposed surface area.

Animals also have some control over the absorption of heat from the sun. Specifically, dark skin and fur absorb more heat than do light skin and fur. Polar bears possess very dark skin, ideal for maximizing the absorption of solar radiation in their Arctic habitat. But what is the use of dark skin if it is covered by a coat of white fur? Well, actually, a polar bear's fur is not white—it only appears white. In fact, each hollow shaft of hair guides the light down to the darkly pigmented skin, where it is absorbed. This design allows polar bears to combine absorption of solar radiation with white-appearing fur: dark fur would make them terribly conspicuous against the white snowy background of their environment, a decided disadvantage when trying to sneak up on their next meal.

EVAPORATION

Recall that conductive and radiative heat transfer occur when one object is warmer than another. These processes can do little to cool us down, however, when the outside world is hotter than we are, because under such conditions heat flows into us, not out of us. Luckily, evaporation plays a vital role on those scorching summer days when conduction and radiation fail us. Indeed, it is the sole natural mechanism of heat exchange that allows us to remain cool even when the air temperature exceeds our body temperature.

Evaporation is the process whereby liquid water changes to water vapor. Physiologists measure the heat of vaporization of water in calories: 580 calories' worth of heat are released when one gram of water is vaporized from the skin of a typical sweating human, thus cooling the skin (and ultimately the body). For this mechanism to work, however, it is necessary that the air above the skin be relatively dry—and the drier the better. This is why we are more prone to heat stroke on humid days: When the air is already saturated with water, we still sweat, but our body water simply drips down our skin without vaporizing. Without evaporation, fluid is lost but body heat is not.

Human sweat is a fluid that is produced by eccrine glands that are located in the outer region of the skin. Relatively few other animals sweat. Most of the ones that do are quite large, such as cattle and camels; horses too are known to "lather up." But sweating is not the only way to take advantage of the benefits of evaporation. Many animals without sweat glands employ a manual strategy. Rats and other rodents possess large salivary glands in their face and neck that release extra saliva into the mouth during heat stress. The animals use this saliva to moisten their forepaws, whereupon they spread the saliva over their snout, head, and body. Some monkeys even urinate on themselves to achieve the same effect.

Another activity that promotes evaporation is panting, which is employed by many small ungulates (hoofed mammals), such as sheep and goats, and by carnivores, such as dogs. Panting entails rapid, shallow breathing over the moist and highly vascularized surfaces of the tongue and the nose, resulting in evaporative cooling of the blood, which then flows back to the brain and body and cools them.

On close examination, panting is a remarkable feat. Dogs, for example, increase their respiratory rate tenfold during panting, from 30–40 breaths per minute to 300–400 breaths per minute. Intermediate values are not found—panting either occurs at full tilt or not at all. More important, dogs achieve this fast respiratory rate with a very low expenditure of energy because their highly elastic respiratory system minimizes the work of breathing. Contrary to popular opinion, humans do not pant—we simply cannot generate the rapid breathing that is required. If we were to try to breathe fast enough to mimic a dog during panting, the heat generated by our respiratory muscles in that prodigious effort would overwhelm any thermal benefit from evaporation.

As already mentioned, evaporation is vital because it is the only way to stay cool when environmental temperature exceeds body temperature. Such conditions often occur in deserts, where the intense heat and the lack of shade during the day present large desert mammals with a clear choice: either use copious amounts of water to cool the body through evaporative cooling and risk dehydration, or save water by forgoing evaporative cooling and allowing body temperature to rise. (A third option is to avoid the daytime heat entirely by building underground burrows and coming out only at night; this nocturnal strategy is adopted by many desert rodents, but it is not available to large mammals such as camels.) In hot environments where water is scarce, the first option is not feasible. Therefore many large desert mammals, including the camel, have adopted the second strategy. These animals simply surrender to the heat during the day, allowing their body temperature to rise well above 40°C (104°F), and then passively dissipate the accumulated heat to the cool nighttime air. It is this ability to "store" heat during the day (rather than an ability to store water, as has often been suggested) that has allowed camels to gain their reputation as the ideal desert ride.

If you have ever encountered a camel, perhaps you noticed that it is not naked but rather has a short coat of hair. And if you were visiting a country where indigenous peoples still employ camels for transportation in hot, arid environments, you may also have noticed that camel riders often travel covered in loose-fitting clothes. Why do both camel and rider dress up for the hot journey? First, fur and clothing not only retard heat loss to a cool environment but also retard heat gain from a hot environment. Second, when a camel or a human sweats, it is most useful if the sweat evaporates slowly at the skin surface, where it can cool the skin and continue to draw heat from the body. Excessive fur and tight clothing are not beneficial because they retard the passage of water vapor from the skin surface, resulting in wet fur or clothing and interfering with the effectiveness of sweating as a heat-loss mechanism.

So there we have the physics of heat exchange: the three primary processes, conduction, radiation, and evaporation, plus one auxiliary process, convection. As we have seen, animals use a variety of behavioral responses and physiological strategies to work around the constraints imposed by the laws of physics. There is, however, one constraint not yet discussed whose effects are felt everywhere, especially with regard to the regulation of body temperature.

THE SURFACE LAW

An elephant and a mouse die simultaneously. The mouse cools down to the temperature of the surrounding air within hours. For the elephant, it takes days.

You place a piece of sweet candy in your mouth. After a minute of patient sucking, you bite into the candy, breaking it into a dozen pieces. You are rewarded with a splendid burst of sweetness.

Many of the most important biological processes take place at surfaces. It is at surfaces that cells and organisms exchange material with the outside world. It is at the surface of a leaf that solar radiation is absorbed to begin the process of photosynthesis. It is at the surfaces of the miniature air sacs (alveoli) in the lungs that oxygen is extracted from the air we breathe. And it is at the surface of the capillary that that same oxygen diffuses from the blood into a cell to help fuel the cell's metabolism.

These three examples—leaves, alveoli, and capillaries—illustrate a fundamental feature of biological organization: the use of branching to efficiently increase the area of biological surfaces. This feature has developed independently in many organisms throughout the biological world. Moreover, this feature is driven by the same physical property that makes elephants cool more slowly than mice and candy explode with sweetness when bitten into small pieces. That property, called the Surface Law, is simply this: *as the volume of a physical object increases, its relative surface area decreases.*

Our intuition may tell us otherwise. After all, a basketball has much more surface area than a Ping-Pong ball. That's where the word *relative* comes in. To illustrate this concept, consider the smallest cube in Figure 2. Let's call this a unit cube, and let's say that the length of each of its sides is 1 centimeter (cm). Therefore the surface area of each side is 1 cm^2, and the total surface area of the cube is $6 \times 1\ cm^2 = 6\ cm^2$. The volume is simply $1\ cm \times 1\ cm \times 1\ cm = 1\ cm^3$. Therefore the cube has a surface-to-volume ratio of $6/1 = 6$.

Now consider the middle cube in Figure 2, which is composed of 8 of the unit cubes. We can determine its surface area by counting its number of exposed sides (24) for a total of 24 cm^2. Because it is made of 8 unit cubes, its volume is simply 8 cm^3. Therefore its surface-to-volume ratio is $24/8 = 3$.

Finally, let's repeat this process for the largest cube in Figure 2. Because this cube comprises 64 unit cubes, we can again determine its surface area by counting its exposed sides (96) for a total of 96 cm^2. And again, because it is made of 64 unit cubes, its volume is simply 64 cm^3. Therefore its surface-to-volume ratio is $96/64 = 1.5$.

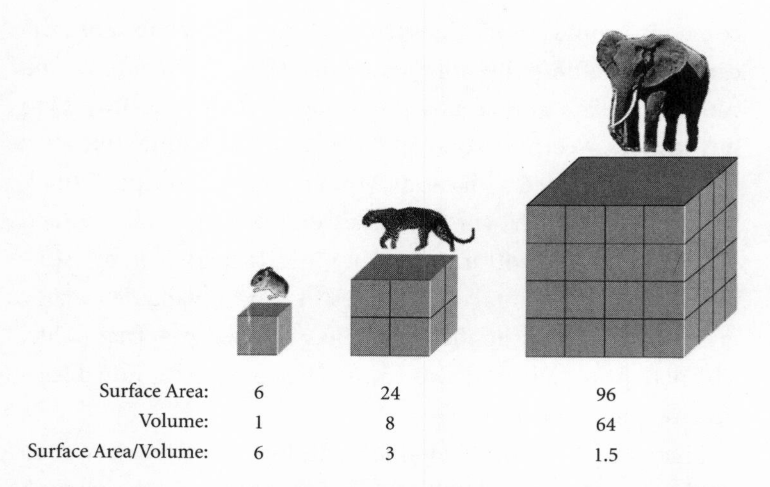

Figure 2 Change in relative surface area as objects increase in size. The cube on the left is a unit cube. As larger cubes are constructed using the unit cube, the surface-to-volume ratio decreases, a relationship referred to as the Surface Law.

We see from this exercise that the surface-to-volume ratio (that is, the relative surface area of the object) decreases as the object gets larger—in this example, from 6 to 3 to 1.5. Thus, as an object grows, its volume grows faster than its surface area. This principle is known as the Surface Law, and it is as fundamental to an understanding of biology as the Laws of Thermodynamics are to an understanding of physics.

Now let's consider that piece of candy that you place in your mouth. When the taste buds on the surface of your tongue come into contact with the surface of the candy, you sense the candy's sweetness. Because you can't taste the inside of the candy, there is a limit, set by the surface area of the candy, as to how much sweetness you can taste at any one time. But after years of experience with candy, you know that a simple crunch between your teeth will release a burst of sweetness. In fact, you are releasing nothing by doing this. Instead, you are simply increasing the total surface area of candy available to your taste buds by breaking the single piece into a dozen fragments. The benefit is increased sweetness. The cost is a rapidly disappearing piece of candy.

Like sweetness, heat is gained and lost at surfaces. A dead elephant cools more slowly than a dead mouse—even though they start at the same body temperature—because the elephant has, relative to its volume, less surface area through which body heat can flow from inside to outside. This basic relationship between size and relative surface area drives many of the differences between mammals of different body sizes.

Mammals as disparate in size as mice, humans, and elephants all maintain a body temperature of 36–38°C (97–100°F). Consider the following analogy: If we open the drain of a pool but wish to maintain a constant water level, we have to fill the pool at a rate that perfectly offsets the amount of water draining out of the pool. But the Surface Law stipulates that the drain of a mouse-sized pool is larger, relative to the pool's size, than the drain of an elephant-sized pool. Consequently, to maintain a constant water level in a mouse-sized pool, we must either add water at a faster rate than we do in an elephant-sized pool or find a way to compensate for the size of the relatively large drain.

Mammals have manipulated the size of the drains through which their bodies lose heat by increasing or decreasing their coats of fur. Consider the amount of fur found on large animals such as elephants and rhinoceroses as compared with the amount of fur found on minks. Indeed, if an elephant were to wear a mink coat, it would rapidly overheat and die of hyperthermia.

Small mammals, however, cannot solve all their heat-loss problems with fur. A mouse with fur as thick as that of a mink would not be able to move under the enormous weight. Therefore mammals have also manipulated the rates at which they *produce* heat. To appreciate this phenomenon, consider that while an elephant has a higher total metabolic rate than a mouse, if we measured the metabolic rate of an elephant-sized pile of mice, the pile of mice would have a higher metabolic rate than the single elephant. In technical terms, we say that the metabolic rate of a mouse *per gram of mouse* is greater than the metabolic rate of an elephant *per gram of elephant.*

The cascade of effects resulting from the Surface Law continues: Because of the mouse's high demand for energy, it must lead a different kind of life than the elephant does.

Unlike the elephant, it cannot spend much of its day sleeping and lounging about; instead it must be constantly on the lookout for its next meal—to support its energy habit. This habit gives mice and other small mammals a distinctive frenetic look that elephants and rhinoceroses simply don't have.

The Surface Law also has an impact on the ultimate variable: longevity. Mice and other small rodents live on the order of a few years; dogs about 10 years; elephants on the order of decades. The high surface-to-volume ratio of small mammals leads to a fast-paced life and, inevitably, to early burnout, while elephants lumber along at a much slower pace. The Etruscan shrew, at three grams one of the smallest mammals (and at 12–18 months one of the shortest lived), possesses a heart that races along at over 1,000 beats each minute, compared to 600 for a mouse, 150 for a dog, and only 30 for an elephant. Interestingly, the reciprocal relation between longevity and pace of life balances out so that the hearts of shrews, mice, dogs, and elephants all beat approximately 800 million times during the very different lives of these species. (Humans reach that number of heartbeats in only twenty-five years, but our allotment of beats, like the size of our brain, is relatively greater for our body size than those of our mammalian cousins.)

Although the easiest way to illustrate the Surface Law is to enlarge an object, such as a cube, without changing its shape, it should be obvious that changes in shape can dramatically affect heat loss. Imagine taking a spherical lump of pasta dough and using all of it to make spaghetti. Clearly, the spherical lump of dough has less total surface area than the accumulated surface area of the many pieces of spaghetti. Accordingly, humans and other animals have adapted their body shapes to their local thermal environments. For example, mammalian species that inhabit hot desert areas have enlarged ears or elongated tails, which increase the amount of skin surface through which heat can be lost. A similar strategy is employed by humans adapted to life in arid environments, as evidenced by the rangy build and lanky limbs of the inhabitants of equatorial Africa. The stout physique of the Inuit peoples inhabiting the Arctic regions of North America represents the other side of this strategy: stoutness entails a reduction in limb length relative to total body size and, therefore, a reduction in heat loss.

In addition to evolutionary processes that modify anatomy and physiology across generations, animals can exhibit, over days, weeks, or months, anatomical and physiological changes in response to new local environments. This process is referred to as acclimation or acclimatization, depending on the cause of the change in environmental conditions. For example, while llamas have evolutionarily adapted to life in the high altitudes of the Andes, people who normally live in a low-altitude location but travel to a high-altitude location will, usually within weeks, acclimate to the low-oxygen conditions. Acclimatization occurs in response to long-term or seasonal changes in temperature, as when a dog grows its "winter coat" each year. Similarly, humans develop increased tolerance (or acclimatize) to cold during the winter months through hormonal and other kinds of adjustments.

As we have seen, the Surface Law places significant physical constraints on some of the most fundamental biological processes. But biology has fought back against these constraints. One way, as mentioned earlier, has been the evolution of branching. A tree trunk gives rise to large branches, each of which gives rise to small branches, and so on, until leaves—which provide the surface area that satisfies the tree's metabolic needs—are found at the end of tiny twigs. The branching is the tree's way of working around the Surface Law, a way to maximize surface area by making room to produce many leaves. The respiratory and circulatory systems of vertebrates also use branching to achieve the same end: the cumulative surface area of all the alveoli in human lungs is roughly equivalent to the size of a tennis court.

Our daily life provides a number of illustrations of the way branching increases surface area. Many of us who suffer with cold hands during the winter have noticed that mittens keep our hands warm better than gloves do. Why? Because even though we do not change the size of our hands by wearing mittens instead of gloves, mittens significantly reduce the amount of surface area exposed to the cold outside air by removing the nooks and crannies between the fingers.

TERMITE NESTS AND THE SURFACE LAW

From my childhood days I can vividly recall horror movies, such as *Tarantula!* and *The Fly*, in which insects are enlarged to human or superhuman proportions. Fortunately for us, no such insect enlargements are actually possible. The reason is simple: human-sized insects would not be able to breathe.

Insects receive oxygen via hollow tubes that wind their way through the body, providing a direct link between the inside of the insect's body and the outside air. If an insect were enlarged, as in *The Fly*, it would suffocate because the small surface-to-volume ratio of the human-sized fly would not be sufficient to provide oxygen to all the cells in its body. Therefore, for the evolution of relatively large animals like lizards, birds, and mammals to be feasible, it was necessary to overcome the problem posed by large bodies. The solution was the evolution of an organ that provides an enormous surface area where exchange of gases can take place—the lung—and a circulatory system by which oxygen can be delivered to all the cells of the body.

Such conflicts between the needs of the animal and the constraints of physics have often provided an impetus for novel evolutionary solutions that, at least in part, resolve the conflict. One of the most exquisite examples of such a resolution comes from the world of the social insects.

Although many of us are familiar with the geometric harmony of the hives of bees, the beautiful designs engineered by African termites are less well known. From the outside, these nests look like little more than conical clumps of dried mud, sometimes reaching a height of sixteen feet. Look inside, however, and these *termitaria* reveal a complex array of compartments and chambers. And with a closer look, one can see that had the design of these nests been the brainchild of a human rather than that of millions of termites guided by a blind evolutionary process, we would be impressed with the human's engineering skills.

Unlike the pesky domestic termites that bring down property values and provide a lucrative living to exterminators, African termites are extremely sensitive to humidity and temperature. When exposed to dry air, these termites die from desiccation, a sensitivity that has resulted in termite nests whose air is nearly fully saturated with water. And when the temperature in a nest drops below the preferred level, termites cease to grow and develop. Because of these dependencies on humidity and temperature, most African termites rarely leave the nest.

The many species of African termites build nests that vary greatly in size, shape, materials, and complexity as well as geographical location. The temperature of termite nests also varies greatly from species to species, in part because thermal stability in the nest is a function of the nest's size (that is, its surface-to-volume ratio) and wall thickness. The species that build large nests (with low surface-to-volume ratios) with thick walls are treated to an internal environment that is largely independent of changes in the temperature of the outside air. In fact, in such species, nest temperature is maintained at the level that tropical termites prefer: 30°C (86°F).

But there is a cost to large nests with thick walls: ventilation is compromised. Termites in the nest breathe oxygen that arrives from the outside air through the surface of the nest, and if the nest is too large, relative surface area decreases such that there is not enough oxygen to go around. Also, inadequate ventilation can lead to a dangerous buildup of carbon dioxide. This conflict between thermal stability and ventilation, mediated by the Surface Law, places a limit on how large a termite nest can become.

At least one species of African termites has evolved a clever nest design that resolves the conflict between thermal stability and ventilation. The nest consists of a basement, a brood chamber, and an attic; the basement is filled with fungus that the termites cultivate so as to produce heat through fermentation. In the attic, channels are dug that pass down the outer wall of the nest and open up in the basement. It is the presence of these channels that is the stroke of genius.

Here's how it works: Beginning in the nest's basement, air is warmed by the fermentation of fungus and by the 2 million termite inhabitants; carbon dioxide is released as a metabolic by-product. The warm nest air rises to the attic, where it encounters the cooler air in the wall channels. As it loses heat, the air falls down through the channels, and as it falls, the high levels of carbon dioxide diffuse through the walls of the nest to the outside air. By the time this air has fallen from the attic to the basement, it is 5°C (9°F) cooler and its carbon dioxide levels have decreased 1.5 percent. (The air, in other words, is circulated by convection just as a rolling boil is produced in a pot of water on a stove.) Therefore, the channel system allows the colony to accomplish what individual termites cannot: growth to human-sized proportions.

HEAT DEATH

We have seen that physical laws and processes constrain and shape the thermoregulatory responses of animals. The Surface Law requires that these responses change dramatically as body sizes change; conduction, radiation, and evaporation, plus convection, provide the ground rules by which animals must abide. To maintain their body temperature, animals have evolved structural adaptations such as fur, physiological adaptations such as sweating, and behavioral responses such as putting on clothing or building nests.

But why regulate body temperature at all? The answer is that failure to keep body temperature within narrow limits is often fatal.

There is no single, ideal temperature for all animals. Among mammals, the monotremes, the group that includes the platypus and the echidna, regulate their body temperature at 30–31°C (86–88°F); the marsupials, the group that includes the kangaroo and the koala bear, regulate their body temperature at 35–36°C (95–97°F); and the placental mammals, the group that includes humans, horses, and dogs, regulate their body temperature at 36–38°C (97–100°F). Among birds, the songbird species (the passerines), such as canaries, regulate at 40–41°C (104–106°F), while species such as chickens and ostriches (the non-passerines) regulate at 39–40°C (102–104°F).

Just as there is diversity in regulated temperatures, there is diversity in lethal temperatures. At one extreme, Antarctic fish that live at a water temperature of –.9°C (30°F; the below-freezing value is correct because the salt in ocean water lowers its freezing point to –2°C) will die in water that is only 6°C (43°F). At the other extreme, the eggs of one species of freshwater crustacean can withstand temperatures within 1°C of boiling. Among mammals and birds, body temperature is typically regulated at 4–6°C (7–11°F) below lethal temperature.

A number of factors are thought to contribute to heat death in animals. One is that many of the enzymatic processes that govern our most fundamental biochemical processes depend critically on temperature. Many of these processes entail complex chains of reactions, and each link in the chain depends on all the others. Therefore, if some enzymatic reactions are affected by changes in temperature more than others, the coordination of the entire chain of reactions can be disrupted. It is as if a candy factory suddenly began adding twice the usual amount of sugar to each vat without doubling the other ingredients. The fact that Antarctic fish and humans, although many of their biochemical processes are similar, have evolved different body temperatures means that the regulation of their biochemical processes must also have adapted to those body temperatures. Thus when Antarctic fish encounter the "hot" water temperature of 6°C (43°F), the chain of enzymatic reactions in their bodies becomes uncoordinated and they die. The same is undoubtedly true for humans during heat stroke.

Another factor that can contribute to heat death is the narrow gap between healthy and dangerous body temperatures. In placental mammals and birds, temperatures of only 41–42°C (106–108°F) can cause brain damage. Many animals experience such high brain temperatures during vigorous activity, exposure to very hot environments, and fever. Humans are particularly susceptible to heat stroke while exercising on a hot humid day when heat produced by the muscles cannot be adequately dissipated; for example, in August 2001 Korey

Stringer, a 340-pound offensive lineman for the Minnesota Vikings, clad in a tight-fitting uniform and "protected" by his helmet while practicing during a heat wave, succumbed to heat stroke and died when his brain and body temperatures soared above 42°C (108°F). The many contrivances that mammals and birds have evolved to protect their brains attest to the necessity of preventing overheating, as we shall see in Chapter 4. And not only brain tissue but nerve, muscle, and blood all behave differently as temperature changes. Clearly, our very survival depends on adequate control of our internal thermal environment.

2 BEHAVE YOUR-SELF

ON THE SOUTHWEST BANK of the Dead Sea, at the lowest point on Earth, a flat-top mountain rises 1,400 feet above the barren ground. On the top of this mesa sprawl the ruins of an ancient fortress, famous for the mass suicide of all but a few of its 960 Jewish inhabitants in 73 CE after a two-year siege by 15,000 soldiers of the Roman legion. Although once a formidable obstacle to the best Roman warriors, this mountain fortress, known as Masada, can now be visited by any tourist who boards the tram that ascends the mountain, rising above the walled perimeters of the Roman encampments still visible below. Alternatively, the fearless tourist may wish to scale the mountain on foot, using the "snake path" that winds its way up the eastern slope. This trek is best done by moonlight in the early morning hours, before the Sun's intense heat discourages even the fittest travelers.

Regardless of how one arrives at the top of Masada, the ruins of the fortress reveal the necessities of a self-sustaining community in an inhospitable environment. For example, water, undoubtedly the most valued commodity at Masada, was gathered and stored using a sophisticated system of aqueducts and cisterns. But Masada was more than just a fortress. Herod the Great, who ruled as king of Judea at the pleasure of the Romans from 37 to 4 BCE, adorned Masada with luxurious palaces, including one that flows in three levels down the northern edge of the mountain, providing a breathtaking view of the Judean Hills to the north and west and the Dead Sea to the east. Herod also constructed on the top of Masada an elaborate series of connecting rooms that reflected the luxurious lifestyle to which he, as a Roman king, had become accustomed. For on the top of this lonely mountain, situated in one of Earth's most desolate locations, scorched and desiccated by the Sun, Herod the Great built a sauna.

To be more accurate, Herod built a Roman bath, or thermae—a set of large chambers, each of which was designed and built to provide an array of thermal sensations to the bather (Figure 3). Although the exact ritual surrounding the thermae is not known with certainty, it is believed that the bather, after

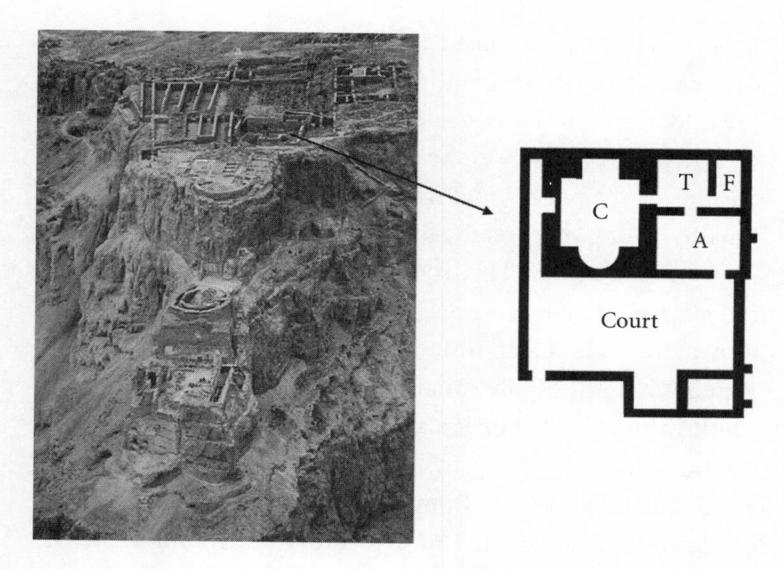

Figure 3 Masada. Left: View of the northern ledge, with terraced palace and, on top, rectangular storehouses. The arrow indicates the large bathhouse, or thermae. Right: The floor plan of the thermae, featuring an entry court, an apodyterium or disrobing room (A), a frigidarium or cold room (F), a tepidarium or warm room (T), and a caldarium or hot room (C).

being anointed with oil and engaging in exercise, disrobed in the *apodyterium* and then entered the *caldarium,* or hot room. At Masada one can still see the remains of the caldarium's raised floor, below which a fire was maintained; this space beneath the floor is called a hypocaust. The heat from the hypocaust was channeled through vertical flues in the wall of the caldarium until it was released near the ceiling.

When the bathers were ready, they would move to the *tepidarium,* a chamber maintained at a warm temperature. Finally, again when ready, they would move to the *frigidarium,* or cold room. This room often contained a pool, and indeed a small one can be seen in the remains of the frigidarium at Masada. Over time, the use of the thermae by the Romans became highly ritualized, designed as a means of promoting bodily cleanliness and relaxation as well as social interaction.

As far as we know, Roman thermae were not explicitly designed to satisfy the thermoregulatory needs of the bathers. We might ask, however, what would happen if a person, unaware of the complex rituals associated with the thermae, were simply locked inside and allowed to move freely from chamber to chamber. What would this person do to achieve thermal comfort?

In fact, we know the answer to this question, although not explicitly for humans in the thermae. Many investigators have used an apparatus that bears a striking resemblance to the thermae to investigate the abilities of lizards and other animals to select thermal environments behaviorally and, by doing so, to regulate their body temperature. Indeed, it was the discovery of this simple apparatus that transformed our understanding of temperature regulation in lizards and other animals.

LOUNGING LIZARDS

What is the relationship between the thermae and the transformation of our understanding of reptilian thermoregulation? First, consider that most of us have been taught to think of reptiles as *cold-blooded.* The validity of this term seemed to be supported by early studies in which a reptile (such as a lizard) was caught in the wild and housed in a cage that lacked a heat source. When investigators measured the animal's body temperature in the laboratory, its value was always close to room temperature. In other words, in comparison with mammals, reptiles were cold-blooded. Moreover, when measurements were made throughout the day, body temperature was found to vary with air temperature, a condition called *poikilothermy* (from the Greek *poikilos,* meaning variegated, and *thermos,* meaning warm).

But these early laboratory observations did not jibe with temperature measurements made in the wild. When an investigator captured a reptile during the day and quickly measured its body temperature, it was often found to be much higher than the air temperature, and body temperatures recorded in the field did not vary nearly as much as those recorded in the laboratory. In the field, reptiles appeared to be neither cold-blooded nor poikilothermic (at least during the day when heat from the Sun was available).

The resolution of this paradox came when investigators began offering reptiles in the laboratory a choice of thermal environments. Like Romans in the thermae, reptiles were now allowed to move freely between a reptile-sized *caldarium* and *frigidarium.* Given the choice, they shuttled back and forth between the two. Shuttling, as we now recognize, enables these animals to regulate their body temperature within a very narrow range, just as they do in the field when they shuttle between sun and shade.

The sensitivity of investigators to the availability of behavioral choices has led to a richer appreciation of the thermoregulatory capabilities of reptiles. Although they are technically categorized as poikilotherms, under the right conditions reptiles can exhibit the high and stable body temperatures that are characteristic of *homeotherms* (from the Greek *homeo,* meaning same, and *thermos,* meaning warm). That is, they use behavior to regulate their body temperature within a narrow range. This is a clear example of how failure to consider the importance of behavior to an animal's physiological well-being can lead investigators terribly astray.

The importance of behavior to physiological regulation is often overlooked, even to this day, and failure to consider it can have disastrous consequences. Perhaps the most famous example of this was provided in 1940 by Lawson Wilkins and Curt Richter of Johns Hopkins University. They reported the story of a three-year-old boy who had been admitted to a hospital in Baltimore because he exhibited accelerated development of the sex organs. He also exhibited an excessive craving for salt that began when he was one year old. The boy's problems were blamed on deficient functioning of the adrenal cortex, an endocrine organ that, among other things, secretes aldosterone, a hormone that helps the body to retain salt. Unfortunately, he was placed on a normal hospital diet, and he died within seven days. As was subsequently appreciated, the boy's craving for salt was a behavioral adjustment to his malfunctioning adrenal cortex. The low-salt hospital diet, and the removal of opportunities to adjust his salt intake behaviorally, killed him.

Presenting reptiles with thermal options that allowed for the full expression of behavior revolutionized our understanding of their thermoregulation. As a result, we no longer speak of animals as cold-blooded and warm-blooded. Nor are the terms *poikilotherm* and *homeotherm* particularly meaningful. Much more useful are the terms *endothermy* (from the Greek *endon,* meaning within, and *thermos,* meaning warm) and *ectothermy* (from *ektos,* meaning outside), which refer respectively to an animal's ability or inability to produce significant quantities of heat internally. Thus reptiles, as ectotherms, rely on external sources of heat for the regulation of body temperature. This reliance places a premium on behavioral, rather than physiological, adjustments.

The internal production of heat is expensive in terms of energy, an investment of resources that attests to the importance of thermoregulation for mammals and birds. But the fact that most animals are ectotherms means that behavior must be the most fundamental of all thermoregulatory mechanisms; in other words, the tendency of animals to orient and move toward sources of heat must have preceded, by many tens of millions of years, the ability to produce heat internally. Indeed, even such seemingly simple ectotherms as lice, maggots, and worms are capable of robust, rapid, and reliable behavioral responses to thermal stimulation.

To study behavioral thermoregulation in lice, we provide them with an environment that presents a wide spectrum of temperatures, from cold to hot, so that any movement brings them either closer to or farther away from their preferred temperature. (The term *preferred* should not be construed as indicating a cognitive decision on the part of the louse. Instead, in this case, *thermal preference* is defined as the environmental temperature at which a louse, when given a choice, settles.) To construct such an environment, we need only heat a strip of metal on one end and cool it on the other end, thereby creating a continuous thermal gradient from one end to the other.

Such an apparatus is sufficient to reveal highly organized behavioral responses. For example, a body-louse placed on the cold end of such a thermal gradient will meander down the gradient, eventually sampling from the entire gradient before restricting its movements to a narrow range of temperatures. The body-louse's behavior need not result from a highly complicated mechanism. A behavioral rule that states "Move when too hot and move when too cold" is enough to produce the thermoregulatory behavior of many organisms.

The apparatus for studying thermoregulatory behavior need not be linear. Radial thermal gradients have been constructed in which the center of a disk is chilled and the temperatures increase radially and continuously from the center to the periphery (Figure 4). Therefore, unlike the linear gradient in which the regions of constant temperature (or isotherms) are organized into strips, the radial gradient produces concentric isotherms. If a nematode worm is placed on such a radial gradient, it moves in a circular motion, restricting its movements to a 0.1°C (0.2°F) circular region of constant temperature. The nematode's ability to perform this feat appears to be due to a pair of sensory organs, called amphids, located in the worm's head region, which provide thermal information to neurons in the worm's brain. In turn, these neurons control the worm's warm-seeking and cold-seeking behavior. When this temperature-sensing system is disabled,

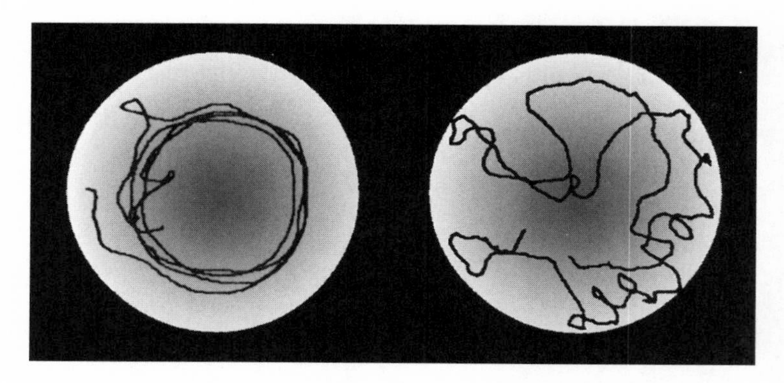

Figure 4 Traced movements of nematode worms on radial thermal gradients. Left: This normal worm is circling the petri dish at 20°C. Right: This worm lacks the neurons that provide sensory information about temperature.

as in the worm whose path is shown on the right in Figure 4, thermoregulatory behavior is severely disrupted. The presence of specific temperature-sensing organs in the worm suggests that its thermoregulatory behavior is considerably more complicated than that of the body-louse.

Over and over again, investigators have demonstrated that such organisms as lice and worms exhibit thermal preferences when given a choice of temperatures. But why do these animals select one temperature rather than another? The answer to this question is simple: temperature influences virtually every biochemical and physiological process in the body, and for each process there exists a range of temperatures at which bodily functions are optimized. Sometimes these optimal temperatures are evolved traits, and sometimes they reflect the life history of the organism. For example, in the nematode worms just discussed, the preferred temperatures correspond to the temperatures at which they were grown in the laboratory.

The thermal gradients constructed by clever investigators in the laboratory are relevant to the real thermal problems faced by animals in the wild. A clear example of this is the gradual cooling of ocean water at increasing depths, which produces a *thermocline*. Each thermal layer within a thermocline is an aquatic roadway for fish, providing a microenvironment that is well suited to their physiological needs. Moreover, each thermal layer is well suited to some species of fish but not to others. Therefore, just as only a small portion of the Earth's surface is navigable by automobile, only a small portion of the ocean's vastness is navigable by each of the millions of fish species that inhabit it.

A LITTLE HELP FROM OUR FRIENDS

We have seen that ectotherms require an external source of heat in order to thermoregulate. Although endotherms need not rely on external sources of heat to regulate body temperature, behavioral thermoregulation is important to them as well. Mammals and birds behaviorally thermoregulate continually throughout the day and night to minimize the amount of energy they must use for the production of internal heat.

Certain types of behavioral thermoregulation are undoubtedly familiar to us all. On a cold day or night, we crawl under the covers, curl up, drink hot chocolate, light a fire, and perhaps set the thermostat in the house to a higher temperature. On a hot day or night, we throw off the covers, sprawl, drink a soda over ice, jump into a pool, and perhaps set the thermostat in the house to a lower temperature. Behavioral thermoregulation is aided by the adaptability of behavior to novel circumstances. When placed in a hot environment, even rats can learn to pull a string to deliver a refreshing shower of cold water.

Sometimes we endotherms depend on the help of others to satisfy our thermoregulatory needs. This dependence is obvious in birds, whose eggs must be incubated at the right temperature for a long time. Although egg incubation is typically a tranquil process in which embryos develop quietly within their elliptical chambers while being carefully tended by the parents, not all avian embryos are so passive. For example, as the embryos of the American white pelican are nearly ready to hatch, they vocalize when their temperature becomes too high or too low. These vocalizations, emerging from inside the egg, stimulate the mother to change her behavior so as to return the embryo's temperature to an acceptable level, whereupon the vocalizations cease.

The thermal interdependence among endotherms is taken to the extreme in Emperor penguins because of the bitter Antarctic cold that they endure for the sake of reproduction. The process begins with the male fattening up for a 115-day period during which he must court a female, mate with her, incubate their egg, and survive without access to food (the nearest ocean meal is 50 kilometers away). When the female lays the egg in mid-May and goes off with the other females to spend the winter at sea (yes, winter; Antarctica is in the southern hemisphere), the males are left behind to incubate the egg for 65 days while enduring 200-kilometer-per-hour winds that can drive the wind chill down to –60°C (–76°F).

Under these incredibly harsh conditions, the male Emperor penguin has two concerns: successfully incubating the egg and surviving the winter. During the incubation period, the egg sits on the male's feet and is covered by a flap of naked abdominal skin that folds over the egg to insulate it. (When the female returns in mid-July, the egg is gently transferred to her feet and the shelter of her abdominal flap, where it soon hatches and the fragile chick remains protected from the still-frigid weather; see Figure 5. At this point, the father is free to return to the open water and enjoy his first meal in four months. Thereafter, the mother and father both participate in

Figure 5 An Emperor penguin chick peeking out from under its mother's abdominal flap.

caring for their chick until, at the tender age of five months, it can fend for itself.)

For his own survival, the male has four layers of feathers as well as a number of circulatory and respiratory adaptations for conserving heat. But even these adaptations are not enough for an individual to survive in this environment. Therefore, hundreds or thousands of male penguins, each wobbling around with an egg couched between its feet and its abdominal flap, stay huddled together throughout the winter, moving in and out of the huddle to satisfy their thermal needs of the moment. Because the surface-to-volume ratio of a huddle of penguins is less than the combined surface-to-volume ratios of all the individual penguins, huddling is an effective thermoregulatory strategy. Indeed, the temperature inside the huddle can rise to as high as 35°C (95°F), even on the coldest of days, and thus heat loss (and energy consumption) by each individual penguin is reduced by 50 percent.

Such collaborations in pursuit of warmth are quite common, as when we snuggle with a lover or, under more primitive conditions, with sled dogs (the so-called five dog night). Some of these collaborations give rise to complex social behavior that serves the basic thermoregulatory needs of the individual animals. Bee swarms expand and contract in warm and cold environments, respectively, thereby maintaining a stable temperature at the center of the swarm. Similarly, the six to eight pups in a typical rat litter engage in a complex and dynamic relationship in which the huddle expands in warm conditions and contracts in cold conditions. Huddles of Emperor penguins exhibit similar behavior. It is important to stress, however, that although these huddles, with their highly organized behavior, suggest the emergence of a superorganism, the group behavior should not be interpreted as being under the control of a "higher" mind. Rather, the complex group behavior emerges as each bee or rat pup or penguin satisfies its own individual needs.

CHANGES IN LATITUDES

Ecology is the study of life on Earth, and perhaps the oldest ecological pattern to have been noticed by early explorers and naturalists is the latitudinal diversity gradient. Credit for the discovery of this pattern—that the warm, humid tropics are filled with more diverse forms of animals and plants than are the temperate and less temperate regions at more northerly and southerly latitudes—has been assigned recently to the geologist and botanist Baron Alexander von Humboldt, who traveled to North America in 1799 under the sponsorship of King Carlos IV of Spain. In 1807 Humboldt published an essay in which he described the relationship between latitude and organic diversity and hypothesized that it is the progressively colder temperatures as one moves away from the tropics that limit the diversity of life at those latitudes.

The strictures imposed by an Earth in which temperatures vary more drastically in the longitudinal direction than in the latitudinal direction are profound but not insurmountable. Unlike flightless penguins, many species of birds have evolved complex navigational and physiological capabilities to make migration possible. Finding one's way from, say, a pond in Canada to an equally small region of northern South America, often at night and over large bodies of water, is no mean feat; nor is it trivial for an animal to perform this feat in the cold, low-oxygen conditions that exist at flight altitude. Nonetheless, birds make these round trips year after year, in response to seasonal temperature changes in order to locate optimal environments for finding food.

The longitudinal migration of birds across the Earth's surface resembles the shuttling behavior of reptiles between two chambers and the movements of Roman bathers in a thermae. The important difference is that the temperature of the apparatus (that is, the surface of the Earth) in which the birds are living changes throughout the year. Therefore, the shuttling of birds between North and South America represents an attempt to maintain constant living conditions in a world in which temperatures fluctuate. An alternative strategy might have been to migrate continuously northward and southward as temperatures fluctuate. But such a strategy would not have satisfied other important ecological needs. For one thing, most birds do not have the luxury of continuous migration because they must establish stable territories for mating, nesting, and raising young.

Like birds, humans have migrated across the Earth's surface in ways that suggest a sensitivity to temperature. In *Guns, Germs, and Steel,* Jared Diamond provides a provocative examination of the factors that have influenced the dissemination of human cultures across the world's continents. He provides many examples, from throughout human history, of rapid transmission of information in a latitudinal direction, in which temperatures vary slightly, and slow transmission in a longitudinal direction, in which temperatures vary dramatically. He notes that plants first domesticated in the Fertile Crescent around 5400 BCE were, only five thousand years later (a relatively short time by historical standards), being farmed to the west in Ireland and to the east in Japan; in contrast, these same domesticated plants found their way down the African continent to Egypt and Ethiopia but were blocked from further progress southward by the Sahara Desert. (When domesticated plants and animals were eventually introduced into southern Africa in the nineteenth century by Europeans, the plants and animals arrived on boats). Similarly, Diamond points out that five thousand years after the llama had been domesticated in the Andes the indigenous cultures to the north in Mexico still had no domesticated animal suitable for work. Diamond offers these and many other examples to illustrate that cultural advances, including the domestication of plants and animals, were more easily transmitted latitudinally than longitudinally throughout human history—in part because of the way thermal gradients and barriers are distributed across the Earth's surface.

The Earth is a complex and varied ecological environment that challenges organisms that attempt to move from one place to another. The movement of domesticated plants has occurred most rapidly across similar latitudes, partly because of the sensitivities of the plants themselves to day length and temperature, but also, no doubt, because of the thermal sensitivities of human farmers, who found it easier to migrate to places that exhibited environmental conditions similar to those of their homelands. For example, in the United States over the last two hundred years, Scandinavians and Eastern Europeans overwhelmingly settled in such states as Minnesota and Wisconsin and avoided such states as Texas and Arizona.

To gain a better appreciation of these human migratory patterns, imagine that you are looking down on a globe of the Earth from above the North Pole and you can observe the increasing temperatures moving southward in concentric circles. If you flatten the globe into a two-dimensional map with the North Pole at the center, the result is similar to the radial thermal gradients used for the study of thermoregulatory behavior in nematodes (Figure 4). Like nematodes in the radial gradient, humans have traversed the globe by following the isothermal paths that circle the globe at each latitude. Initially, when all such traversals were done on foot, the travelers followed these paths very closely, just as many fish remain within isothermal layers in the water. But more recently, as the ease and speed of transportation have increased with the advent of boat travel and then flight, we have been freed from these ancient constraints and have come to resemble birds, flying northward and southward in our long-distance migrations from one hospitable locale to another.

There will come a time, however, when the hospitality provided by our earthly host will come to an end. Our Sun's luminosity has increased 30–40 percent over the past 4.6 billion years, and the additional 10 percent increase in luminosity that will occur over the next billion will probably be enough to overheat our planet. If our descendants wish to preserve the Earth as a safe haven for life in our solar system, it may be possible, through intricate maneuvering, to use the gravitational energy of a large asteroid to nudge the Earth's orbit outward just enough to increase its distance from the Sun and thereby maintain Earth's temperature at a suitable level. This maneuver, devised by the astrophysicist Donald Korycansky and his colleagues, could be repeated every few thousand years to compensate for the steadily increasing temperature of the Sun. Such a fantastic yet feasible cosmic act of behavioral thermoregulation would lengthen our planet's usefulness to us by many thousands of years until, with the final dying of the Sun, we would be forced to leave our ancient shelter in search of another. In the meantime, enjoy the weather.

3 THEN BAKE AT 98.6°F FOR 400,000 MINUTES

GENES HAVE STOLEN the hearts of America's youth. When I ask my students what the essence of a human being is, they give me the answer they have been taught to give. The essence of a human being, they say, is the genotype—our genetic make-up. Similarly, if I ask them what the essential difference between a man and a woman is, they focus on the genes, although this time they can be more specific, providing textbook answers: females have two X chromosomes while males have one X and one Y. Of course, some students insist that the essential difference is the external genitalia, and others provide a more functional distinction (males inseminate and females gestate), but the winner is clear. Genes, they say, define gender.

Let's ask the same question about a chocolate cake: What is its essence? Is it the eggs, the flour, or the cocoa? No, those are only the ingredients. Okay, what about the pan, the mixer, or the oven? No, those are only the instruments used to make the cake. Perhaps the real essence of the cake is the recipe, because the recipe contains the instructions that allow any of us to make the cake regardless of where we live, where we purchase the ingredients, or what kind of oven we have (although baking times do vary with altitude). The recipe is unchanging and reliable: it contains the *information* regarding how to make our cake.

One often reads that the genes provide the recipe, blueprint, or program for making a human being. But in fact genes do nothing of the kind. Genes contribute to the making of proteins. That's it. There is no blueprint to a human being, there is no program, there is no recipe. Human beings, like all organic beings, are constructed through a complex, bi-directional interaction between the genes and the environment in which they are embedded. A recipe is not a cake, and how and when the ingredients are combined is just as important as acquiring the right ingredients in the first place.

We are well aware of the importance of baking for the creation of a cake, but the role that temperature plays in the growth and development of animal and human fetuses is rarely explored. As we will see, even a small, brief increase in temperature during gestation can have devastating effects. In addition, recently discovered roles of temperature in the development of sex and thermal preference in reptiles provide clear examples of the way our seduction by genes continues to cloud our understanding of development.

THERMAL MUTANTS

How can there be no recipe or blueprint for animal development? How can such a complex organism as a human or a dog develop without a set of instructions? This is the very issue that disturbed so many people when Charles Darwin proposed his theory of evolution. Before Darwin, the Argument from Design had held sway among those interested in the natural world. That argument had been invoked by such thinkers as Aristotle and Newton, as well as by William Paley in the early nineteenth century, and is still repeated today by so-called Creation Scientists. Paley's famous rendition is this: if we found a rock and a watch lying in a creek bed, only the watch, with its obvious complexity, would lead us to conclude that it had been planned by a human designer; by analogy, human and animal complexity must have been designed—by God. Darwin's contribution was to propose a mechanism, natural selection, that is capable of producing organismal complexity without the need for a designer.

As an example of a complex process that occurs without a blueprint or a set of instructions, consider a pot of water sitting on a stove. At first the stove is turned off and the water molecules are largely inactive and are distributed freely throughout the pot. Now we turn on the stove, and the molecules move faster and faster as they are heated. Finally the water comes to a rolling boil. Simply by turning on the stove, we have replaced disorder (the random distribution of water molecules) with order (the rolling boil, in which the water molecules rise and fall in a circular pattern within the pot). The question is: Where is the recipe, blueprint, or program for the rolling boil?

There is no recipe, blueprint, or program. The rolling boil *emerges* from the complex interactions of the parts, that is, the pot, the water, and the heating element. And there is one more crucial ingredient—gravity. A rolling boil occurs because heat is provided from below and there is cool air above, creating a thermal gradient from bottom to top. The rolling boil, then, depends upon the presence of a gravitational field in which hot water rises and cool water falls. If we try to identify the cause or the essence of the rolling boil, or to discern the blueprint for it, we will be stymied. There is no single cause, no essence, and no blueprint. *There are only the parts and their interactions.*

Boiling water is a simple example, but the lesson is important. Sometimes, when investigating a complex system that has recently changed in some way, we attribute the cause of this change to a single part of the system. If a baseball team loses year after year but then suddenly has a winning season after the addition of a single player, we may be tempted to say that the cause of the team's newfound success is the presence of the new player, all the while disregarding how the performances of the other players have changed. We may even ignore the fact that the newly added player had arrived from another team that had also lost year after year. Again, individual parts and their interactions are of equal importance.

Every cell in your body—skin cell, brain cell, liver cell, heart muscle cell—has the same DNA residing within its nucleus. And yet each of these cell types is vastly different. Obviously, genes do not determine the fate of a cell. Rather, genes are differentially activated or suppressed during the development of a fetus depending upon the local environment in which the cell is embedded, resulting in differential expression in cells that will eventually perform a specific function within a specific organ, be it skin, brain, liver, or heart. This process is highly dynamic and interdependent, with a variety of environmental factors playing roles in shaping gene expression throughout the organism. But there is a factor during development that is so all-encompassing that we seldom think about it. This factor is the temperature within the womb.

Demonstrating the importance of gestational temperature to development is easy. For example, investigators exposed pregnant rats to an environment warm enough to cause a

3.5°C (6°F) increase in body temperature for a mere thirty minutes on the ninth day of gestation (rat pups are born on the twenty-second day of gestation). When examined on the eighteenth day of gestation, the fetuses exhibited hideous developmental abnormalities, including facial clefting and maldevelopment of the eyes and brain. Some of the fetuses even lacked a brain altogether. That such a brief exposure to so small an increase in body temperature, one that is well within the range of a fever, can cause these severe malformations is extraordinary. And it must be stressed that the genes themselves were unaffected by the mother rat's brief exposure to a warm environment; what changed was the expression of the genes.

Because it was a change in gestational temperature that resulted in the deformed rat fetuses, we are tempted to say that increased temperature *caused* the developmental defects, even though we are disinclined to say that a normal gestational temperature *causes* the development of a normal rat fetus. Such biases toward assigning causation to the changing variable are common. Consider this everyday example: a boy is throwing a ball around with a friend. What causes the ball, once it leaves his hand, to follow the path that it does toward his friend? A reasonable answer is that the force provided by the boy's arm is the cause. But we must not forget about the Earth's gravity. Without gravity, the ball would continue on a straight path over and beyond his friend's head. We have all learned to throw balls in a gravitational environment, providing a force on the ball that allows it to reach its intended destination. It is easy to forget about gravity because it's always there. Similarly, genes have evolved to "expect" a particular thermal environment, and when that thermal environment changes, so too will the activities of genes.

SEX WITHOUT SEX CHROMOSOMES

Adults treat baby boys and baby girls—or what they think are baby boys and baby girls—very differently. When told that an infant is a boy, they talk about how tough "he" is, using a tone of voice and handling the infant in accordance with "his" toughness. But when told that an infant is a girl, adults handle "her" gently, speak to "her" softly, and admire "her" delicacy. The degree to which parents reinforce these gender stereotypes with their own infants predicts how their children will behave later in life.

Although we all recognize these cultural sources of gender identity, we also recognize the more fundamental biological reality. But where does that reality come from? Is it true, as most of my students believe, that gender has an essence that resides within the X and Y chromosomes? To answer this question, let's begin by looking at what these chromosomes do.

Mammalian embryos begin the developmental process in a neutral sexual state. Although the embryo's genes may be XX or

XY, it does not yet possess ovaries or testes. Rather, it has what are called *ovotestes,* organs that have the capability of developing into ovaries or testes. Whether the ovotestes go one way or the other depends upon the presence or absence of a hormone that is produced only by the Y chromosome. If this hormone is present, then the ovotestes will develop into testes; if not, they will develop into ovaries. If testes develop, the production of testosterone (and other related hormones) by the testes will guide the embryo down a path toward maleness. But the cascade of events along that path can be disrupted at any stage, resulting in strange outcomes. For example, if an XY "male" embryo is injected at the right time with a drug that blocks the transition of the ovotestes into testes, a normal female will be produced, although she will have both X and Y chromosomes.

Although there are some interesting examples of anomalous gender development (in Androgen Insensitivity Syndrome, for instance, XY embryos have a mutation that renders them insensitive to testosterone, resulting in a child raised as a female, only to discover at puberty, when menstruation does not begin, that she has two testicles in her abdomen), such occurrences are rare. Under normal circumstances, the X and Y chromosomes determine at some fundamental level whether an embryo will develop into a male or a female. But this raises an interesting question: Are sex chromosomes used by all animals to produce gender differences?

The answer is a resounding *No!* There are many examples of animals that have no sex chromosomes whatsoever and yet manage to develop distinct sexes. Consider certain species of coral reef fish that live in schools made up of many females along with one male, which is responsible for inseminating all the females. We can distinguish males from females because males have testes and females have ovaries. In addition, each sex exhibits behaviors and coloration that are indicative of maleness or femaleness. On occasion, the school's single male dies, leaving a school composed only of females. But this unisexual condition does not last long because the dominant female in the school, which is usually the largest female, quickly changes her sex to male. When she does this, the change is complete: her ovaries convert to testes and she begins behaving like a male in all respects, including the ability to inseminate the other females in the school.

What prompts her to make the switch? The testes of male coral reef fish produce testosterone, and the males excrete the hormone in their urine. Because schools are tightly knit communities, swimming close together, the presence of testosterone in the local water is sensed by the female fish and inhibits them from switching sex. But when the male dies and the concentration of testosterone in the water decreases, the dominant female switches sex more quickly than the others, and this newly transformed male begins producing testosterone and excreting it into the water, thereby inhibiting the others from also switching sex. Because each fish has the intrinsic capability to become either sex, coral reef fish are referred to as *sequential hermaphrodites.*

Such gender bending in coral reef fish demonstrates that sex chromosomes are not needed to produce functional sex differences. But there are other examples that illustrate the arbitrariness of the mammalian reliance on sex chromosomes. Alligators, crocodiles, and many lizards and turtles lack sex chromosomes, but nonetheless they hatch either as male or as female. Something happens during incubation of the eggs that determines sex. The mechanism is not chromosomal—it is thermal.

The process is called temperature-dependent sex determination (TSD), and it works differently in different species. In alligators, eggs hatched at low incubation temperatures become males, while eggs hatched at high incubation temperatures become females; intermediate temperatures produce clutches comprising both males and females. In some turtles, the pattern is exactly the opposite. And in crocodiles, low and high incubation temperatures produce females and the intermediate temperatures produce males.

Through a process that scientists do not yet understand, different incubation temperatures activate a hormone that influences the development of ovotestes into ovaries or testes. What is striking, however, is the similarity between the processes by which reptiles and mammals develop sex. For both groups, the critical turning point—whether to develop ovaries or testes—is the same; the only difference is the mechanism that leads to that turning point. Incubation temperature does for a reptile what sex chromosomes do for a mammal. Clearly, in mammalian species that gestate their young internally and that maintain a relatively constant body temperature, temperature-dependent sex determination would not work; therefore, for mammals, the advent of sex chromosomes was a reasonable solution.

DEVELOPING A PREFERENCE

Why do you and I regulate our body temperature at approximately 37°C (98.6°F) and not, say, at 34°C or 40°C? It is no coincidence that this is the body temperature of our parents and grandparents; thus it is quite obvious that our body temperature is an inherited characteristic. But what is the mechanism of this inheritance? It could be that there are genes that govern the development of body temperature in our species; this is what we usually think of when we think of inherited characteristics.

But is it possible that our species-typical body temperature of approximately 37°C derives from the fact that our mothers not only share their genes with us but also provide a secure and comfortable developmental environment whose temperature is regulated at 37°C throughout the 400,000 minutes of gestation? Although this question is fundamental to our understanding of the development and evolution of thermoregulation, it has not received much attention. Certainly one impediment to investigation of this problem in mammals, especially placental mammals such as rats (and humans), is the difficulty of manipulating the gestational temperature of the fetus without grossly affecting its development or influencing the mother's behavior, either of which would make interpretation of the results difficult. Therefore, one possible research strategy for testing this *epigenetic* hypothesis—the hypothesis that the early developmental environment interacts with the genes to establish regulated body temperature—is to use a species in which manipulation of the developmental environment is easy. Enter eggs.

Species that lay eggs—including birds and many reptiles—offer the opportunity to examine the effect of manipulating the embryonic environment on post-hatching behavior and physiology. Indeed, thanks to the efforts of many investigators, we now know that incubation temperature, in addition to determining sex in some species, influences a variety of developmental and adult features, including the amount of time from laying to hatching (incubation time), skin pigmentation, and running speed. But our question here is somewhat more complicated: Does incubation temperature modulate the level at which body temperature is regulated and, therefore, the functioning of the brain?

Building on an earlier study using crocodiles that suggested an effect of incubation temperature on post-hatching behavioral thermoregulation, my colleagues and I conducted an experiment using a species of gecko, the Madagascar ground gecko. This species satisfied certain important criteria. Its embryos tolerate a wide range of incubation temperatures, and they do not exhibit temperature-dependent sex determination; the latter criterion is particularly important to ensure that any effect of incubation temperature on behavioral thermoregulation is not merely due to a difference between the two sexes. For this experiment, gecko eggs were randomly assigned to incubators whose temperatures were maintained at 23°C (73°F), 26°C (79°F), or 30°C (86°F). When the eggs hatched, the hatchlings were immediately tested in an apparatus in which they could shuttle between a cold (16°C, 61°F) and a hot (41°C, 106°F) surface. Using an infrared camera mounted above the apparatus, we were able to monitor the hatchlings' behavior and temperature as they shuttled from the cold side to the warm side and back again over a period of many hours.

The results of this experiment were unequivocal: incubation temperature had a systematic effect on the *preferred temperature* of the hatchlings. The strongest and most reliable finding was that hatchlings that had been incubated at 23°C shuttled off of the cold side of the apparatus at lower body temperatures than did hatchlings incubated at 30°C, while hatchlings incubated at 26°C exhibited intermediate responses. In other words, just as temperature modulates the development of gender, it also modulates the development of the thermoregulatory system.

We do not yet know what the implications of these findings are for the lives these geckos lead or for the evolution of body temperature in reptiles and other animals, including humans. As stated earlier, very little work has been done on this problem. But the work that has been done has been promising. For example, it has been shown that incubation temperatures in the Muscovy duck (which is, like all birds, an endotherm) have an impact on post-hatching thermoregulatory capabilities, suggesting that our findings in reptiles may not be limited to ectotherms.

Although extending these findings to mammals may be difficult, it is not impossible. Let's imagine for the moment that investigators were to find that altering gestational temperature in a mammal influences the behavioral and physiological responses of the newborn infant and that these effects are stable throughout the animal's lifetime. What would be the implications of such a finding? First, it would suggest that, within limits, our thermoregulatory system is established during gestation through an epigenetic process—a bi-directional interaction between genes and environment. Second, it would suggest that while there may be advantages to regulating our body temperature at the particular value that we do, the evolutionary process by which we arrived at that temperature did not merely involve selection for one or more "temperature genes."

Finally, because many drugs of abuse influence body temperature, such a finding would suggest that we should begin examining the health consequences of altered gestational temperatures for human infants. For example, ecstasy (or MDMA) has been in common use for nearly two decades and is now popular with thousands of people around the world, especially young people who consider it to be a harmless weekend or party drug. In its pure form, ecstasy enhances the activity of the neurotransmitter serotonin in the nervous system, resulting in a variety of psychological and physiological effects. Ecstasy has been shown to inhibit normal thermoregulatory responding in rats such that their body temperature falls if they are housed in a cold room and rises in a hot room. Such passive thermoregulatory responding becomes particularly dangerous for humans when the amphetamine-like qualities of ecstasy kick in and users engage in vigorous activity for hours on end, as typically occurs at dance clubs and raves. The combination of abnormal thermoregulatory functioning and prolonged exercise can result in profound hyperthermia, damaging the brain, the liver, and other organs. In fact, one of the leading causes of death due to ecstasy is heat stroke. One has to wonder how many young children are now suffering the consequences of having been exposed to unnaturally high temperatures during gestation because of the use of ecstasy and other drugs with powerful thermoregulatory effects.

That both sex determination and behavioral thermoregulation are influenced by incubation temperature highlights the fact that genes do not possess privileged causal status in the developmental process. Genes are part of the machinery, and this machinery includes all the parts and their interactions. Sometimes the trigger for a developmental event comes from outside the skin and sometimes from inside the skin. It doesn't really matter. What does matter from an evolutionary standpoint is that the resulting organism works.

4 EVERYTHING IN ITS PLACE

THE HISTORY OF THE MEASUREMENT of body temperature has been, to put it bluntly, a search for the most convenient orifice. When I was a small child and my mother suspected that I was "coming down with something," there was one preferred method. My mother, having gathered preliminary data by applying the back of her hand to my forehead ("Let me feel your *keppi,*" she would say, using the Yiddish word for head), would conclude that more accuracy was required. Lying in bed, I would listen anxiously as she rifled through the bathroom cabinet. Upon returning with the thermometer, she would command me to turn over.

Eventually, to my relief, oral temperature replaced rectal temperature as my mother's favorite. And since that time a variety of alternative methods have hit the market, including an infrared device that measures the temperature of the inner ear—quickly, with reasonable accuracy, and without embarrassment.

Which orifice provides access to the "real" body temperature? This question is more difficult to answer than one might think. As is so often the case in science, the methods used shape the questions asked and the conclusions drawn. As we will see, an animal's body has many body temperatures, each of which has its own functional significance for the animal.

COLD FEET IN WADING BIRDS

Imagine a bird wading in a shallow lake, searching for its next meal. The water is very cold, and the bird has to stand there for hours, remaining as still as possible to avoid alarming its potential prey. Each moment, blood flows through the bird's long spindly legs and is cooled by the water, then flows back up to its body and mixes with the rest of its blood. In time, this cooled blood has a significant impact on deep body temperature, forcing the bird to increase its internal heat production to maintain a constant temperature. As a consequence, it consumes additional energy, thus adding to its need for food—a need that motivated the bird to enter the lake in the first place.

Wading birds have evolved a simple yet effective mechanism to deal with this problem. This mechanism, called *countercurrent heat exchange,* requires nothing more than bringing the artery and vein in each leg into close physical proximity. When the two vessels are juxtaposed in this way, the warm arterial blood leaving the body pre-warms the cool venous blood returning from the feet. By the time the venous blood enters the body cavity, it has been warmed sufficiently that the bird need not increase heat production to maintain body temperature. Because it merely requires the anatomical alignment of the two blood vessels, countercurrent heat exchange is a very cheap solution. It does, however, have one conspicuous cost—very cold feet.

WARMING UP TO INSECTS

For thermoregulatory biologists working with uncooperative nonhuman subjects, there were two traditional methods for measuring body temperature. In the first method, a noose was used to subdue an unsuspecting creature (for example, a lizard) and a temperature-measurement device was inserted into the cloaca (similar to the rectum in mammals). In the second method, the favorite for the measurement of body temperature in insects, the temperature-measurement device was placed inside a syringe needle, which was then inserted into the subject's abdomen.

For decades the study of insect thermoregulation was dominated by the view that insects were, for all intents and purposes, hardly more active in controlling their body temperature than pet rocks. Technically speaking, these animals were considered poikilotherms, animals whose body temperature swings up and down at the whim of air temperature. Study after study documented this "fact": insect body temperatures never exhibited significant deviations from the surrounding thermal environment. This conclusion, however, was based on studies in which the standard location for the measurement of body temperature was the abdomen. After all, how could it possibly matter where you measured the temperature of an insect that weighed only 1 gram?

As it turned out, it did matter where the insect's temperature was measured. When the zoologist Bernd Heinrich, in the 1970s, began investigating the thermoregulatory abilities of the sphinx moth, he was surprised to find that these animals do indeed regulate body temperature. Rather than relying on abdominal temperature, Heinrich measured the temperature of the thorax, the compartment of the insect's body that contains the heart and flight muscles (in mammals, reptiles, and other vertebrates, the thorax is the cavity enclosed by the ribs). His finding was revolutionary: the thorax of a moth is considerably warmer than the rest of the moth's body. In other words, the thorax is where the thermal action is.

Why would the thoracic temperature of a sphinx moth, or any other insect, be higher than its abdominal temperature? The thorax contains the flight muscles, and, as we know from watching athletes warming up before a race, warm muscles function better than cold muscles. In fact, moths and bees engage in pre-flight warm-ups, in which the flight muscles that control the wings' upstroke and those which control the downstroke fire in near synchrony, producing heat through shivering while minimizing movement of the wings. Upon attainment of a muscle temperature that can support flight, the shivering ceases and the insect takes off. Also, like wading birds, these insects use a countercurrent heat exchanger: cool blood flowing from the abdomen to the thorax is preheated by warm blood flowing from the thorax to the abdomen, thereby maintaining thoracic temperature as much as 25°C (45°F) higher than abdominal temperature (and thus highlighting the folly of judging insect thermoregulation solely by measuring abdominal temperature). This is a remarkable feat for such tiny animals—and it seems all the more remarkable because we once considered them incapable of even rudimentary thermoregulation.

HOTHOUSE FLOWERS

Perhaps even more surprising than a thermoregulating moth is a thermoregulating flower. But for the last thirty years the zoologist Roger Seymour and a number of other investigators have been studying the abilities of certain flowers to generate heat. They have recorded flower temperatures as much as 34°C

(61°F) above the temperature of the surrounding air. One species of arum lily, the skunk cabbage, is even known to melt snow. What is more, some plants—including the skunk cabbage and the sacred lotus—regulate their heat production with precision similar to that of mammals and birds.

A convincing demonstration of this regulation is provided by an experiment with the sacred lotus (Figure 6). The floral chamber of the sacred lotus contains a central receptacle, which itself contains the female reproductive organs, or *stigmas.* Surrounding the receptacle are the rod-like male reproductive organs, or *stamens.* To study thermoregulation in this flower, investigators measured receptacle temperature over the course of six days as the petals opened and pollen was shed. Prior to the opening of the flower, receptacle temperature increased and decreased in lock step with air temperature, but as the flower opened and entered a reproductive state, receptacle temperature was maintained at 30°C (86°F), even as air temperature dipped as low as 10°C (50°F). This maintenance of receptacle temperature appears to be an example of true regulation, because changes in heat production during this period mirrored changes in air temperature. Finally, when the petals opened fully and the pollen had been shed, the period of thermoregulation was over and receptacle temperature fluctuated once again with air temperature.

Figure 6 The sacred lotus, a thermoregulating plant found in Asia and Australia. The many stamens, which produce pollen, are visible encircling the centrally located receptacle, which contains the seeds. The receptacle is responsible for the majority of heat production in this flower.

Why would a flower thermoregulate? Although a few hypotheses have been proposed, the relationship between the period of thermoregulation and the flower's reproductive state suggests the most intriguing possibility—that the floral heat is an attractant to the flying insects that pollinate the flower and carry its pollen to other flowers. For example, in one species of philodendron, thermoregulation and floral opening commence when the female reproductive organs are ready to be pollinated; the flower's heat provides a warm invitation to beetles to fly inside. Soon after, the flower ceases its heat production and closes, trapping the beetles inside, where for the next twelve hours they pollinate the flower's female reproductive organs and are themselves coated with pollen from the flower's male reproductive organs. Then the flower reopens, and the beetles are free to fly off and pollinate another flower.

SAVE THE SPERM!

Near the beginning of his first term, President Bill Clinton was widely ridiculed for appearing on MTV and answering the following question: Does the President wear boxers or briefs? Missing from the question was a third option—bikini underwear. Bikini underwear is still around, but there is good reason for it not to be, at least for men. The reason? Bikini underwear kills sperm.

Like the thoracic muscles of the sphinx moth, the testes of mammals, especially placental mammals like rats, rams, and humans, have special thermal needs. The abdominal temperature of mammals is higher than the ideal temperature for the production of healthy, viable sperm; moreover, high temperature in the testes can result in genetic mutations and testicular damage. Therefore, as regulation of a high body temperature evolved, it became advantageous for some animals to find a new home for the testes outside the abdominal cavity. Hence the evolution of the scrotum.

The scrotum is a highly sophisticated thermoregulatory device by which testicular temperature is maintained 2–9°C (4–16°F) below body temperature, depending on the species and the thermal environment. With the scrotum we encounter yet another countercurrent heat exchanger. Blood within the veins of the scrotum is cooled by the surrounding air; the pendular construction of the scrotum makes such cooling possible. As this venous blood flows back to the body, the proximity of the veins to arteries entering the scrotum allows for the cooling of warm arterial blood arriving from the abdomen. Thus the arterial blood is pre-cooled before it reaches the scrotum, just as blood is pre-warmed in the legs of wading birds.

The longer the distance over which the artery and the vein are aligned with one another, the more efficient will be the pre-cooling. Now, let's say a man falls into Lake Michigan in February, thus threatening his testes (and his entire body for that matter) with extreme cooling. Sensory receptors on the scrotal skin detect the change in temperature and elicit a reflexive contraction of the tunica dartos muscles, thereby pulling the testes closer to the abdomen and preventing excessive pre-cooling of the arterial blood coming from the abdomen. On hot days, the opposite occurs: the scrotal muscles relax, increasing the distance between the testes and the abdomen and allowing for greater pre-cooling of the arterial blood. And now we see the problem with bikini underwear: by keeping the scrotum close to the body it prevents cooling of the testes, resulting in a real threat to the production and viability of sperm.

A study of rams provided a compelling demonstration of the body's commitment to maintaining testicular temperature. A special chamber was constructed that allowed the investigator to manipulate the temperature of a ram's scrotum. When scrotal temperature was increased using this device, the ram began panting, a response that helps to decrease body temperature and, in turn, scrotal temperature. In addition, the scrotum began sweating, a response that is complementary to panting and that, through evaporative cooling of the scrotal skin, helps to reduce testicular temperature.

As just described, a ram with an artificially warmed scrotum begins to pant to drive down body temperature. But what happens when, despite this panting and decreased body temperature, scrotal temperature is unaffected because the scrotum is encased in a warm chamber? Interestingly, the ram continues panting as long as its scrotum is overheated, even if its body temperature decreases more than 2°C (4°F). It appears that the ram is more committed to maintaining the temperature of its testes than to maintaining its deep body temperature.

FISH EYES

The ocean is a complex thermal environment. As noted in Chapter 2, ocean temperatures decrease with increasing depth along a thermal gradient called a thermocline. The layers of the thermocline are like highways extending over great distances, providing aquatic avenues for fish to navigate through their underwater world. Because most fish cannot produce significant quantities of heat internally, they are dependent on the temperature of the water in which they are swimming. And because fish function better at some temperatures than at others, they are not free to explore the entire ocean; their world is considerably smaller than the expansiveness of the ocean suggests.

While some fish ride the straight and narrow, others are the aquatic equivalent of off-roaders. These large pelagic (open water) predators, like the best-known predators on land and in the air, rely on a keen visual sense. But unlike the body temperatures of lions and hawks, the body temperature of even large fish depends predominantly on the temperature of the surrounding environment—that is, water. Why does this matter? It matters because the eyes and brain work better when they are stable and warm: as a general rule, the neurons that make up the eyes and brain fire faster and require less time to recover between firings when they are warm. In other words, the integrative functions of the nervous system are highly dependent on temperature.

Imagine a bluefin tuna, chasing small fast-swimming prey as they lead it toward and away from the surface. As the tuna tries to keep the prey in sight, struggling to move its seven-hundred-pound body, it can experience changes in water temperature of 15°C (27°F) in very short periods of time. If the tuna were like most fish, the temperature of its eyes and brain would seesaw up and down with the water temperature. But it is different.

As a large fish with a relatively high metabolism, the tuna produces plenty of heat. The challenge is to retain as much of this heat as possible. To do this, tuna use an arrangement of arteries and veins similar to that used by wading birds, retaining heat in eyes and brain while allowing most of the rest of the body to stay cool (tuna use similar arrangements to maintain high muscle temperatures). This situation, however, is somewhat different from that of wading birds in that the tuna aim to retain the heat produced by the metabolically active eyes and brain. While in birds the arterial blood coming from the body is warmer than the cool blood in the feet, in tuna the cool arterial blood coming from the body must be warmed as it flows to the eyes and brain. To increase the effectiveness of the tuna's system, the major arteries serving the eyes and brain break up into many small branches, thereby increasing the surface area available for countercurrent heat exchange. We use the term *rete,* which means net, to describe the branching of a large arterial vessel into smaller vessels. Veins carrying warm blood from the eyes and brain envelop the small arteries of the rete and pre-warm arterial blood on its way to the eyes and brain. Using this countercurrent system, bluefin tuna can maintain eye and brain temperatures more than 6°C (11°F) above water temperature and, perhaps more important, buffer the neural machinery within eyes and brain from the rapid changes in water temperature that these fish experience as they live out their predatory lives.

Some species of shark have evolved a slightly more sophisticated system for maintaining eye and brain temperatures. These sharks, called the Lamnid sharks, include the porbeagle and the mako. Like tuna, these species possess a rete, but instead of merely anatomically juxtaposing arteries and veins, the shark's rete is surrounded by a large venous cavity (called a venous sinus) that literally bathes the two hundred small arterial branches of the rete in the warm venous blood returning from the eye and brain. By the time the small arteries in the rete rejoin to form larger vessels carrying oxygenated blood to the eyes and brain, the blood has been warmed sufficiently to maintain eye and brain temperatures 5°C (9°F) above that of the surrounding water.

We have seen with tuna and sharks that much can be done by using clever and complex vascular arrangements to retain the heat produced within the eyes and brain. Although a difference in temperature of 5°C may seem small, it may well make the difference between catching and losing one's prey. But some fish have gone one step further, moving beyond the passive retention of heat produced as a by-product of neural metabolism to the active production of heat to maintain brain temperature. These fish, known collectively as the billfishes, include the marlin and the sailfish. Their creation is the brain-heater tissue, an organ comprising muscle that has been modified for heat production. The brain-heater tissue's size tells us a lot about its importance: in one 136-kilogram swordfish with a 2.2-gram brain, the heater tissue weighed 150 grams, nearly 70 times heavier than the brain. In the head of the swordfish, the brain, especially the part of the brain that processes visual information, sits atop the heater tissue and white fat surrounds the brain on the top and on the sides. Therefore, heat produced by the heater tissue is effectively distributed to the brain and the eyes to maintain visual acuity. And this heat is further conserved with the help of a rete that, in swordfish, has as many as twelve hundred arterial branches.

This sophisticated setup reflects the demanding lives of the swordfish and its relatives. Swordfish experience rapid and dramatic changes in their aquatic environment, including daily excursions as deep as 600 meters and changes in water temperature of as much as 19°C (34°F). Experiencing such variable conditions while hunting fast-swimming prey has apparently placed a premium on a system that can buffer the eyes and brain against thermal fluctuations that could disrupt the functioning of the central nervous system.

BIRD BRAINS

Birds live on the thermal precipice: they have the highest body and brain temperatures of all animals, often exceeding 40°C (104°F) at rest. Why? Birds (the ones that fly) strike a balance between two needs. First, they need to be pre-warmed at all times so that instantaneous flight is possible, if necessary. As we have seen with insects, flight requires warmed muscles—and many birds have evolved very high body temperatures. Second, they need to protect their brains from heat damage, which begins at temperatures above 41°C (106°F). Therefore, birds face the challenge of maintaining high muscle temperatures without incurring brain damage. Moreover, because brain tissue is more sensitive to overheating than muscle and other types of tissue, keeping the brain cool allows birds to fly for longer periods than would otherwise be possible.

To maintain a high body temperature without damaging their brain, many birds (including ducks, pigeons, ravens, and roadrunners) have evolved a variety of mechanisms that allow them to selectively protect the brain from overheating. First, birds, like sharks and billfishes, possess a rete system that helps them keep the brain cooler than the body. Recall that in fish the venous blood that is used for countercurrent heat exchange arrives from the eyes and brain, where it has been *warmed.* In contrast, in birds, the venous blood bathing the rete arrives from the face, where it has been *cooled.*

Birds cool the venous blood in their face in a number of ways, one of which is panting: for birds as for dogs and rams, panting is a form of rapid respiration in which air is moved quickly over the evaporative surfaces of the nose and mouth, resulting in the cooling of venous blood. Blood cooled in this way then flows back to the rete, where cooling of the small arteries in the rete takes place by countercurrent heat exchange.

Maintaining a cool brain is particularly important during flight, when the continuously high metabolic rates produce hyperthermia and pose a constant thermal threat to the brain. Birds counter this threat by taking advantage of a natural cooling opportunity presented by flight itself—wind. Studying pigeons during flight, investigators have found that air movement over the cornea cools venous blood in the eye through conduction and convection; this blood then flows back to the rete, where it cools arterial blood on its way to the brain.

LET COOLER HEADS PREVAIL

Like birds, mammals need to prevent their brains from overheating. Humans don't have a rete mechanism at the base of their brain, like birds and billfishes, but many other mammals do, especially the artiodactyls (the even-toed ungulates, a group that includes sheep and gazelles) and the carnivores (the meat-eaters, a group that includes lions and wolves). In these animals, blood flows to the brain through the carotid arteries, but before reaching the brain these arteries branch out into a rete. The blood in this network of vessels flows through a venous sinus that, as in sharks, is a cavity in which cool venous blood flowing back from the mouth, the nose, and the rest of the head bathes the thin arterial vessels of the rete. As these arterial vessels exit the sinus, they join to form a large ring-shaped vessel called the Circle of Willis, which supplies the brain with cool arterial blood.

In mammals that have a rete, venous blood is typically cooled within the nose and the mouth by means of panting. In some animals, including the dog, the nasal cavity has evolved a complex arrangement of bones called turbinates, a scroll-like arrangement that vastly increases the amount of surface available for cooling. Thus, even though dogs have relatively unimpressive retes, they are able to keep their brain more than 1°C (2°F) cooler than the body during heavy exercise.

To appreciate the functional significance of selective brain cooling, consider lions and gazelles. Gazelles are choice prey for lions. During a long chase on the open savanna on a hot day, the gazelle's chances of escaping are determined by its ability to outrun the lion. During a run, the gazelle's body temperature increases as the muscles in the legs and elsewhere produce heat. Blood warmed by the muscles flows throughout the body, increasing the temperature of every organ, including the brain. When the brain reaches a critical temperature that could initiate irreversible damage, the gazelle does not have the luxury of calling time out to allow its brain to cool down—this is not a game. So, to survive, the gazelle must keep its brain cool *during* the run. To do so, it evaporatively cools blood flowing through its nose and mouth, which then flows back to its well-developed rete and cools blood flowing to the brain. In this way, the gazelle keeps its brain temperature more than 4°C (7°F) below its body temperature. The lion, of course, must keep up with the gazelle, and it appears that evolutionary pressures have provided lions with a similar ability to cool their brains.

Many mammalian species, including humans, do not have a fully developed rete but are nonetheless capable of some degree of brain cooling. These animals cool venous blood in a number of ways. Rabbits have large and highly vascularized ears that are effective cooling surfaces, as do African elephants, which flap their ears to increase heat loss through convection. In humans during hyperthermia, venous blood is cooled as it flows through dilated vessels in the face and scalp (this cooled blood protects the brain from overheating either by cooling arterial blood flowing to the brain or, perhaps, by cooling the brain directly). The dilation of the blood vessels in the face is what accounts for the flushing that we experience on hot days, and for that irritant to the easily embarrassed—blushing.

Although we now have a great deal of insight into the diversity of mechanisms by which mammals and birds cool their brains, we are not beyond the occasional surprise. Most recently, the surprise involved a new discovery in horses, which are capable of cooling their brains during exercise even though they lack the mechanisms found in other mammals. What horses (and other odd-toed ungulates, the so-called perissodactyls) do have are two air-filled pouches, located on each side of the head just behind the jaw, which envelop the pair of carotid arteries that supply the brain with blood. By measuring temperature in a pouch and at various sites along the carotid artery, investigators discovered that the arterial blood is cooled as it flows through the pouch during trotting and cantering. How the blood is cooled in these pouches is not yet understood.

The body is a complex thermal landscape in which animals with different needs shunt the commodity called heat to the areas that need it most. In some cases, however, heat is a liability, and predators and prey engage in the thermoregulatory equivalent of an arms race, with each species evolving more efficient and effective mechanisms for cooling the brain. In these evolutionary disputes, it is the cooler head that prevails.

A SENSITIVE BRAIN

It has been suggested that the phenomena that scientists study sometimes reveal their personal weaknesses. My beloved hot tub, which sits behind me as I write these lines, is a testament to the accuracy of that suggestion, for it is a symbol of my poor thermoregulatory abilities. But regardless of one's reason for immersing oneself in hot water, most people clearly prefer to do it on a cool evening rather than on a hot day, or in winter rather than in summer, even though the water temperature doesn't change. Why? The most obvious factor that changes between winter and summer is the air temperature surrounding the part of the body that stays above the water surface—the head. What makes hot-tubbing enjoyable is providing therapeutic warmth to your aching muscles for as long as possible. What limits your ability to stay in the tub is your brain temperature. On cool evenings, the air cools your head sufficiently to keep your brain at a comfortable temperature while your body benefits from the heat. A similar phenomenon occurs on exercise bikes that are designed to generate a breeze that blows on your face: you can exercise longer if your brain remains cool, just as the gazelle, by cooling its brain, can delay the lion's next meal.

Given the sensitivity of brain tissue to overheating, it makes sense that it possesses detectors that help it to protect itself. These thermally sensitive neurons, called thermosensors, allow the brain of humans and other animals to detect changes in its temperature and to organize appropriate responses. Early and compelling evidence that such thermosensors play a role in coordinating both physiological and behavioral responses was provided in 1964 by the biopsychologist Evelyn Satinoff. Satinoff bent a small metal tube into the shape of a hairpin and implanted it into the hypothalamus of an adult rat. (The hypothalamus is a small part of the brain, located just above the roof of the mouth, that plays an important role in temperature regulation.) By circulating water of different temperatures through this tube, she could change the temperature of the hypothalamus without affecting the rat's body temperature. When its hypothalamus was cooled, the rat exhibited both physiological and behavioral heat-gain responses: it shivered and pressed a lever that turned on a heat lamp. The precise integration of the rat's behavioral and physiological responses was beautifully demonstrated by the fact that rats attained the same hypothalamic temperature regardless of whether or not they had access to the heat-providing lever—if they did not, they had to rely on shivering alone.

After decades of research, we now know that there are two broad classes of neurons that function as thermosensors—"warm receptors" that increase their rate of firing in response to an *increase* in temperature and "cold receptors" that increase their rate of firing in response to a *decrease* in temperature.

These two types of thermosensor, found in the brain and spinal cord, in conjunction with thermal information provided by sensory input from the skin and other parts of the body, participate in organizing our physiological and behavioral responses to changes in our thermal environment. It is the integration of these multiple sources of information by the brain that ultimately triggers a behavioral decision to change the setting on a household thermostat, to throw off a blanket, to jump into a cold pool, to turn on a fan, or to get out of the hot tub.

5

COLD NEW WORLD

THE SUDDEN AND UNEXPECTED death of an apparently healthy baby is the leading cause of infant mortality during the first year after birth in the industrialized world. In 1992 in the United States, 6,000–7,000 infants died of Sudden Infant Death Syndrome (SIDS), 95 percent of these deaths occurring between the ages of two and four months. Because SIDS is a diagnosis of last resort, a label applied to *unexplained* infant death, it has not been easy to isolate its cause and to determine how to prevent it. It is not surprising, therefore, that although numerous theories regarding SIDS have been entertained over the years, we still do not know why these infants die. This uncertainty only adds to the self-blame and shock experienced by parents whose baby has died without warning.

When faced with a rare and mysterious syndrome such as SIDS, one looks for patterns in different countries, cultures, and ethnic groups. For example, rates of SIDS are high in New Zealand (especially among the Polynesian Maoris) but low in Hong Kong and Scandinavia. In the United States, SIDS is low among Asian Americans but high among Native Americans and low-income African Americans. Once these patterns are recognized, one can look for behavioral and environmental factors that differ among these groups that might account for differences in rates of sudden infant death.

On the basis of these kinds of epidemiological analyses, two factors have received a lot of attention. First, investigators noticed that the infant's sleeping position—face up or face down—differs between ethnic groups in a way that suggests a contributing role to SIDS. For years, parents in the United States and elsewhere had been told that placing an infant in the face-up position could lead to suffocation if the infant were to vomit. This opinion changed during the 1990s, partly because of cross-cultural evidence that SIDS is less prevalent where infants are put to bed on their backs (and that these infants are *not* at increased risk for suffocation due to vomiting). Such evidence has led to the "Back-to-Sleep" campaign, which advises parents to put infants to bed on their backs, and the results have been dramatic: in the United States, according to the National Institute of Child Health and Human Development, the percentage of infants sleeping on their stomachs dropped from 70 percent to 21 percent between 1992 and 1997, and the number of SIDS deaths dropped to fewer than 3,000 per year, a 50 percent reduction.

One benefit a human infant derives from sleeping on its back is that, if necessary, it can cool down more readily because heat is more easily lost from the chest and abdomen than from the back. This is important, because overheating is one of the characteristic postmortem signs of sudden death in infants. In some cases this overheating is extraordinary, as in one infant whose body temperature was recorded as 42°C (108°F) *five hours after the estimated time of death.* Such overheating could be due to increased heat production (as occurs with infection and fever) or to decreased heat loss, and both of these factors have been implicated in SIDS. The role played by decreased heat loss is particularly interesting because it is so easily correctable. Many parents, especially during cold winter months when SIDS deaths are prevalent, become very concerned about their infant's ability to keep warm during the night. They may overdress and overwrap their child in preparation for sleep, and may also place the child in a warm room, sometimes very close to a radiator. These well-meaning precautions severely limit the infant's ability to use either physiological or behavioral methods to cool down, a predicament that can quickly lead to hyperthermia and death. Perhaps the strongest evidence that overheating is indeed a contributor to SIDS comes from the rare occasions of simultaneous sudden death among twins. In the four such cases examined, the twins' environments were found to be overheated.

The overwrapping of infants in warm rooms is a modern overreaction to an ancient and legitimate concern. The concern is legitimate because the primary threat to most infant mammals' normal growth and development is excessive cold, not excessive heat. Any baby mammal that must devote considerable energy to staying warm will have less energy left over to grow. As we will see, the threat of cold to the developing infant has driven the evolution of physiological and behavioral mechanisms without which the infant could not survive to become an independent adult.

A COOL INTRODUCTION

Growth is work, but so is birth—for infant and mother alike. For placental mammals like humans, birth is an onslaught of adjustment: from fluid-filled to air-filled lungs; from darkness to light; from warmth to cold. The birth transition is especially difficult for those infants born in a relatively immature state of development, such as humans, dogs, and rats. Just to survive birth and its severe challenge of getting enough oxygen, these immature infants possess the ability to summon a rapid and explosive hormonal response that keeps them breathing in conditions that would kill them just a few weeks or months later. The response involves the adrenal gland and the hormones adrenaline and noradrenaline (today more typically called epinephrine and norepinephrine). These hormones are also released by the little-understood and developmentally transient organ of Zuckerkandl. Together, the hormones released by the adrenal glands and the organ of Zuckerkandl help to clear the lungs of fluid, stimulate breathing, and keep the heart beating during the physiologically stressful period that follows birth.

The cells of infant mammals also appear to be better able than the cells of adults to survive severe reductions in the availability of oxygen, a condition called *hypoxia.* Thus a combination of hormonal stimulation and cellular resiliency appears to provide infants with an advantage over adults when the oxygen content of air falls below the normal value of 21 percent. This advantage is only temporary, an adaptation that helps infants to survive challenges that are unique to their stage of development. Such temporary adaptations are known as *ontogenetic adaptations* (*ontogeny* simply refers to an organism's development). Other examples of ontogenetic adaptations are the umbilical cord and the ability to suckle; each serves a vital but temporary function during development.

Many of us have heard reports of children who have fallen into lakes at near-freezing temperatures, sometimes remaining underwater for hours, and have not only survived but made complete recoveries. Similar stories are often heard of children wandering out into the cold and surviving despite severe hypothermia. One such incident occurred in March 2001, when a thirteen-month-old girl named Erika, wearing nothing but a diaper and a T-shirt, wandered out of her house into the frigid night in Edmonton, Alberta, Canada, while her mother and sister were asleep inside. By the time her mother found her, anywhere from thirty minutes to four hours later, Erika was nearly frozen. In fact, she was so stiff that emergency medical workers had difficulty puncturing her skin with a syringe needle or opening her mouth. Nonetheless, after intensive rewarming procedures, Erika's heart began beating at least two hours after it had stopped. That same month, a two-year-old boy named Les, from a farm near Eau Claire, Wisconsin, strayed into the subzero cold in the middle of the night and was not found until four hours later, when his body temperature had fallen to 18°C (64°F). Both children appear to have recovered completely.

Although Les's and Erika's stories of survival are remarkable, they are not altogether surprising. Researchers who perform surgery on infant mammals have found that cooling pups, by immersing them up to their neck in ice water for a few minutes, is a very effective method of anesthesia. Indeed, in many ways this method is preferable to other forms of anesthesia, such as ether. Infants anesthetized using this freezing method can be rewarmed in an incubator with no harmful side effects. This technique can be used with infant rats only until they are approximately ten days old, a developmental shift that has not yet been explained (a similar shift occurs much later in humans, but its timing and abruptness are still unknown).

Whether hypoxia is induced by freezing or by prolonged submersion in water, animals appear to be protected by the cold (we don't hear stories of people surviving for hours without a heartbeat during the summer). The reason temperature makes such a huge difference for survival is that hypoxia is only a problem when there is a mismatch between the supply of oxygen and the need for oxygen. When the supply of oxygen is gone, as when a person is under water, then reducing the need for oxygen is the only way out. Because all the cells of the body have a reduced need for oxygen when they are cooled, extreme cooling actually protects the animal—infant or adult—from dying. This helps to explain how hibernating mammals can allow their body temperature to plummet to near-freezing levels during the winter. It also explains the near-universal response of ectotherms—from bacteria to lizards—to hypoxia: they behaviorally select cooler environmental temperatures to reduce their need for oxygen.

The basic principle that cooling can protect an organism's cells from hypoxia and subsequent death has now been discovered by doctors treating patients who have suffered a stroke. Stroke occurs when a blood vessel in the brain bursts, thereby depriving brain cells of the oxygen once provided by the ruptured vessel. Because brain cells are especially sensitive to hypoxia and will die within minutes of oxygen deprivation, strokes can have devastating and irreversible effects on brain function. It now appears that decreasing the brain temperature of stroke patients can significantly improve their chances of survival and reduce the extent of brain damage. This finding is exactly as one would expect from the compelling testimony of organisms as diverse as bacteria, turtles, hibernating squirrels, and infant mammals.

Given the susceptibility of infants to overheating, as in some cases of SIDS, and their ability to tolerate extremes of cold, one might reasonably conclude that it is overheating that presents the real danger. In a sense this is true. But the fact that infants possess more effective mechanisms for combating cold than heat provides a clue that dealing with cold has been the more important evolutionary challenge. As we will see, this contention is supported by the fact that most mammalian infants possess a specialized organ that helps them to maintain a stable internal environment and thereby promotes successful growth and development.

A WARM HEART

Identifying the nature of animal heat was one of the great quests of early science, beginning with Greek philosophers who identified *innate heat* as the source of life itself. According to Plato and many others, the heart is the source of animal heat and it is the function of the lungs, through the inhalation of cool air, to prevent overheating. Lest one be tempted to view this theory as admirably mechanistic but simply wrong in the details, it must be stressed that for these early thinkers the heat produced by the heart was not a worldly heat but rather was thought to derive from the quintessence, the fifth element (after earth, air, water, and earthly fire itself) of which the heavenly bodies are composed. These views persisted virtually unchanged until the seventeenth century, when René Descartes introduced his mechanistic philosophy with earthly fire as a central player. In addition, under the influence of Descartes, William Harvey, and others, it was soon recognized that many physiological processes could be explained in terms of chemistry; this recognition opened up the possibility of viewing animal heat as a form of chemical combustion. This revolutionary insight, combined with the rapid development of accurate thermometers in the seventeenth century, set the stage for the demystification of the nature and source of animal heat.

We have come a long way in our understanding of how animals produce, control, and distribute heat. As we saw in Chapter 4, animals have not one but many body temperatures, each with its own functional significance. We saw that different animals protect the temperature of different parts of the body—thoracic temperature in moths, testicular temperature in mammals, eye and brain temperature in fish, brain temperature in mammals and birds. The lesson was that if one wants to make meaningful statements about thermoregulation in any given species, one must put some thought into the functional relevance of the temperature that is to be measured. But this lesson does not apply only to the study of different species. It also applies to the study of animals of different ages.

Mammals have adopted a variety of reproductive strategies to maximize their evolutionary success. At one extreme there are animals such as horses and sheep that typically give birth to one offspring at a time; their offspring are precocial, that is, born in a developmentally mature state with fully functional eyes and ears, fur, and advanced locomotor abilities. At the other extreme there are animals such as dogs and rats that give birth to many young simultaneously; their young are altricial, that is, born in a developmentally immature state with closed eyelids and ear canals, naked skin, and limited locomotor abilities. Human infants are more difficult to categorize because they exhibit a combination of altricial and precocial features: they have immature locomotor abilities but mature senses of sight and hearing. From a thermoregulatory standpoint they are similar to infant rats in that they are small (relative to adults) with naked skin, features that limit their ability to retain heat.

As immature mammals with large surface-to-volume ratios, naked skin, little subcutaneous fat, and limited motor abilities, altricial infants, and human infants, appear to have neither the physiological nor the behavioral capacity to keep warm. How do they survive? First, because some altricial infants, such as puppies and kittens, are born with many siblings with whom they can huddle, each infant's individual problem of surface area is offset to some degree. Second, mothers provide heat and insulation to their offspring, both by sharing body warmth and by supplying warm milk for the infants to consume. Finally, the infants of many mammalian species, including humans, possess a specialized organ of heat production—brown adipose tissue or BAT—that provides the infant with much-needed heat when it is cold, including shortly after birth. (Heat production by BAT appears to be a substitute for shivering, the second major form of heat production, which, for reasons that are not yet clear, is less common in altricial infants.) BAT is morphologically similar to white adipose tissue (or WAT, better known simply as fat), but it appears brown because of a high concentration of energy-producing organelles called mitochondria within the fat cells; in addition, the mitochondria within BAT are special in that they have been biochemically modified for the production of heat.

The presence of a mechanism for producing heat that is ready to go at birth raises the question of whether BAT can produce heat in the womb and thereby threaten the mother with overheating. For example, BAT is activated by adrenal hormones; if the mother were to become stressed and her adrenal gland were to begin releasing adrenaline, might BAT begin producing heat and create thermal problems for the mother? Fortunately, this does not appear to be possible. Research on fetal sheep indicates that the placenta produces a hormone that suppresses heat production by BAT. Once the infant is born and its connection with the placenta is severed, BAT is freed from this inhibition and can produce heat.

If infants such as rats and humans are so limited in their ability to retain heat, then why do they go to the trouble of producing heat at all? One answer, first suggested thirty years ago and recently confirmed in my laboratory, is that infant rats direct their heat to the parts of their bodies that need it most. BAT is ideally located, in both humans and rats, for the delivery of warm blood to the vital organs within the thorax, that is, the heart and the lungs (noting the distribution of BAT around the upper torso, one investigator called it a thermal blanket). Keeping the heart warm is particularly important because, like all muscles, it is profoundly influenced by temperature: when warm the heart beats quickly, but when cool it beats slowly. As it turns out, rat pups do an adequate job of producing heat, directing that heat toward the heart, and retaining it within the thorax, while allowing other parts of the body, such as the abdomen, to cool. Thus rat pups and insects such as moths and bees share the strategy of compartmentalizing their heat in order to subserve a specific function—maintenance of cardiac function in mammals and warming of flight muscles in insects.

The production of heat by BAT and its compartmentalization within the thorax are very significant for the rat pups' ability to withstand a range of moderately cold air temperatures. What does *withstand* mean in this statement? Physiologically, it means that cardiovascular and respiratory function are maintained within acceptable limits. Behaviorally, it means that the pups remain asleep. In other words, these small, naked creatures, once thought to be incapable of meaningfully regulating their temperature, exhibit all of the signs of successful thermoregulation.

But there is a limit to how much heat a pup can produce. So what happens when that limit is reached?

BLOOD AND TEARS

Regardless of how old we are or how fit, there are limits to our physiological, behavioral, and cognitive capabilities. We can run only so far and so fast and for so long. We can jump only so high. We can add only so many numbers in our head. We can withstand only so much pain, so much hunger, so much thirst. We can withstand only so much cold. And so it is with rats.

As we have seen, infant rats do an admirable job of thermoregulating as air temperature decreases through a range that I will refer to as *moderate cooling.* There comes a point, however, where the infant can no longer sufficiently increase heat production using BAT to offset its heat loss. I will call the range of temperatures below this point *extreme cooling.* Once this point is reached, heat loss overwhelms heat gain. The result is a precipitous fall in body temperature.

As mentioned earlier, hibernating mammals allow body temperature to fall to near-freezing levels. Nonhibernating animals experiencing such a precipitous fall in body temperature respond vigorously by triggering compensatory physiological reflexes. For example, they stimulate their cardiovascular and respiratory systems to maintain an adequate flow of oxygenated blood to vital organs and muscles. But in hibernation these compensatory responses are not engaged. Rather, hibernators go gently into a near-freezing state. (When it is time to emerge from hibernation, they rely on BAT to raise their body temperatures. They can't warm up by shivering because the rapid muscular contractions that characterize shivering are not possible at such cold temperatures.)

During exposure to extreme cold, rat pups experience many of the same physiological changes as do hibernators, but the rats do not go gently. Rather, they struggle mightily to maintain cardiovascular and respiratory function in the face of rapid cooling. Their first problem comes from the direct effect that hypothermia has on cardiac function: as the heart muscle gets progressively colder, both the rapidity and the strength of its contractions diminish. These changes have a significant impact on the pup's ability to deliver oxygenated blood to its body. A second problem for the hypothermic pup involves the blood itself. Like many substances (such as fat and motor oil), blood becomes thicker, or more viscous, as it gets colder. And as the blood gets thicker, it becomes harder to pump through the blood vessels.

The effect of hypothermia on both heart and blood suggests that the animal will feel distressed, and the behavior of hypothermic pups is consistent with this diagnosis. Whereas at moderate air temperatures pups remain asleep even as their BAT produces increasing amounts of heat, at extreme air temperatures they wake up and begin moving around, as if they are searching for their mother. Furthermore, these pups emit a cry from their tiny throats at a frequency so high that humans cannot hear it. The mother rat does hear this cry, emitted at 40 kHz (most humans cannot hear above 20 kHz), and she behaves appropriately—searching for her pup, then grasping it by the nape of the neck and returning it to the warmth of the nest.

For many years, this story of the infant rat calling to its mother for help was simply accepted at face value. Indeed, many researchers still interpret the pup's cry as an expression of emotional distress and believe that it may provide a model to help us understand the emotional needs of human infants during separation and discomfort. But there is more to this story.

Recall that profound physiological changes accompany extreme cooling in infant rats, including decreased cardiac function and thickening of the blood. These physiological changes might well cause emotional distress; after all, heart attacks are not exactly enjoyable experiences. So it would seem reasonable to view the rat pup's cry as a response to the discomfort brought on by a failing cardiovascular system.

But now consider a sneeze. When the mucous membrane lining our nose is irritated, we reflexively engage in an explosive maneuver that entails the forced expulsion of air through the nose that removes the irritant. In doing so, we produce a sound, and this sound often elicits responses from other people—*Gesundheit* or *God bless you.* A sneeze from a small child may induce a mother to hand the child a tissue. But here's the critical point: we don't conclude from the mother's gesture that the child sneezed *in order to* induce her to hand over a tissue.

The moral of this story is that we cannot infer the cause of a behavior (sneezing) from the response that it evokes (delivering a tissue). This simple point transforms the way we view the crying rat pup and its mother's response. That the mother rat retrieves the pup to the nest when it cries does not necessarily mean that the pup cries in order to get its mother to retrieve it. But why else would the pup cry?

Once again we return to the effects of hypothermia on the heart and the blood. As the cooling heart is less and less able to pump blood into the arteries because the cooled blood is becoming more viscous, the now-thick venous blood pools in the large peripheral veins rather than returning to the heart.

Maintaining blood flow back to the heart is absolutely essential, but solutions are limited. Returning to the warm nest is an ideal solution, and we have seen that the mother will retrieve the pup when she hears the cry. But regardless of the mother's behavior, there is one thing the pup can do on its own behalf: it can contract its abdominal muscles while simultaneously closing its larynx, thus squeezing peripheral blood back to the heart; this maneuver is called the abdominal compression reflex. In fact, this is what we believe is going on when the pup cries—it rapidly contracts its abdominal muscles and, as with a sneeze, incidentally produces a sound. Accordingly, we believe the cry is simply an acoustic by-product of a physiological process that serves to maintain cardiovascular function when times are bad. The mother, for her part, also acts on her own evolutionary behalf by monitoring the physiological condition of her pups.

Human infants also emit sounds that are known to be acoustic by-products of other processes. For example, infants with respiratory problems, especially infants born prematurely, emit an audible grunt when they are in respiratory distress. These grunts accompany a maneuver that is very similar to the infant rat's: by breathing against a closed larynx, the infant produces backpressure within the lungs that expands the alveoli and enhances the infant's ability to breathe, incidentally producing the grunt. This maneuver is known as laryngeal braking because the larynx is used to brake the movement of air as the infant exhales. Laryngeal braking has received considerable attention from investigators, and understanding this maneuver has helped in the treatment of both premature and full-term infants in respiratory distress. In contrast, the abdominal compression reflex has received very little attention. After what we have learned about rat pups, I wonder whether the latter reflex is more important for human infants than has been appreciated thus far.

WORDS OF COMFORT

Mammals are noteworthy for the degree to which their social interactions involve close body contact. We have seen that huddling among littermates is one of the earliest social experiences of altricial infants. Even more fundamental to all mammals is the close body contact that derives from the mammalian way of feeding young, that is, the attachment of the infant either continuously (as in kangaroos) or intermittently (as in humans) to a maternal nipple. Because suckling necessarily entails close physical proximity to the mother, the infant also receives heat from the mother's chest and abdomen.

A mother animal's behavior is closely tied to the thermal capabilities of her young. For example, because infant rats can produce heat internally and retain much of their heat by huddling, the mother is free to leave the nest on occasion and forage for food. In contrast, Syrian golden hamsters are born without the ability to produce heat, leaving the mother as the sole source of warmth; as a result, a pregnant hamster stockpiles food around her nest so that she can be continuously available to her young for at least the first week after she gives birth.

Polar bear mothers face one of the most difficult tasks among mammals—bearing and raising offspring in the Arctic. One might think that polar bears would give birth during the summer, and perhaps produce young that are reasonably well developed and ready to deal with the harsh environment. But in fact polar bears give birth in December, when air temperatures are as low as –40°C (–40°F), to cubs that are very immature—small, blind, and almost completely naked. Even more surprising, the cubs have little subcutaneous fat and their ability to produce heat appears negligible. How do they survive?

The female polar bear builds a snow den, big enough for her and her cubs, whose internal temperature probably hovers around 0°C (32°F). In the den the mother curls up with her cubs, providing them with the insulation they lack. Her milk is high in fat, thus giving the cubs the fuel they need to maintain their relatively high resting metabolic rate. Cared for in this way, the cubs emerge from the den three months after birth, now weighing about 10 kilograms (more than ten times what they weighed at birth), fully furred, and able to maintain a constant body temperature at air temperatures of –30°C (–22°F) without even having to increase heat production. Despite their rapid development, the cubs continue to depend upon their mother for many months for warmth and nutrition (Figure 7).

Human infants also must contend with thermal problems, and, like rats, they produce heat using BAT. Unlike rats, however, humans and other primates are generally born as singletons, which means that they have no siblings with which to huddle. Primates solved this problem by creating a close thermal bond between mother and child. In humans the ancient history of this bond can be seen in the palmar reflex, the grasping reflex that is evoked by placing one's finger in a baby's palm. This reflex helps the infants of our primate cousins to grasp their mother's fur as she goes about her daily business. (Although human mothers no longer have fur, the palmar reflex lives on, like goose bumps, as a vestigial response.) Doctors and nurses are becoming increasingly aware of the need

Figure 7 A polar bear mother and her cub.

for close bodily contact between human mothers and their infants. This form of care, referred to as *kangaroo care,* is an effective way of transferring warmth to the infant and may have a number of important additional benefits.

Given the paramount importance of maternal contact for the provisioning of warmth to mammalian infants, it is not surprising that our language has come to reflect these early thermal needs. It seems likely that the infant's dependence on maternal warmth gave rise to the linguistic pattern in which *warm* typically denotes physical closeness and *cold* typically denotes physical distance. Consider the children's game in which one player hides a target object and the other player searches for it while the child who hid it tells the searcher that she is freezing, cold, hot, very hot, and finally boiling as she moves progressively closer to the target. An implicit assumption of this game, which some call *hide the thimble,* is that the target (or its location) is a heat source.

This is not to say that we always link warm with good and cold with bad. Although we may long to sit in front of a hot fire on a cold winter's night, we have a similar longing to jump in a cold pool on a hot summer's day. Hot fires and cool pools give us chances to regulate our body temperature behaviorally, and cooling down when we feel hot is just as pleasurable to us as warming up when we feel cold. But this observation does not refute the notion that our earliest experiences with temperature are asymmetrical and that *warm* has a generally positive and comforting connotation that *cold* does not.

It is not difficult to think of examples of such uses of thermal language. The oxymoron *cold comfort* is used to describe a comfort that really isn't. We prefer a *warm embrace* to a *cold shoulder.* And such relations are not limited to English. In French, Italian, and German, as well as in tropical languages, warmth is also associated with closeness and cold with distance. The French refer to a warm welcome as *un accueil chaleureux,* Italians as *una dimostrazione calorosa,* and when Germans feel comfortable or at home with someone, they say *bei jemandem warm werden.* On the cool side of the dichotomy, a chilly smile is called *un sourire froid* by the French, a chilly reception *una accoglienza fredda* by the Italians—and a cynical person is referred to in German as *kaltlächelnd,* which literally means "cold smiling."

Languages, like organisms, evolve. They have common ancestors and distant relatives. Therefore, when considering whether the association between warmth and physical closeness is a universal feature of human languages, one must take into account the historical relations among the languages sampled. English, French, and German are closely related languages, so common associations between warmth and closeness in those languages are not evidence of universality. And although some contemporary tropical languages also exhibit the warmth-closeness association, it is difficult to draw sound conclusions when the historical relations among the languages are not considered.

To get a better handle on these issues, Jeffrey Alberts and Gyula Decsy of Indiana University examined two groups of languages that have not shared a common ancestor for many thousands of years: Indo-European, which includes English, French, and German; and Proto-Uralic, which includes Finnish and Hungarian. The same pattern emerged in all the languages they examined: warmth carries with it an affective connotation of closeness and, well, warmth.

Alberts and Decsy traced the etymological roots of *warm* in a variety of Indo-European and Proto-Uralic languages and found that those roots had one of two sources. In some cases, *warm* derived from explicitly thermal words such as *embers* or *fire,* while in other cases, the roots were related to both *warm* and *breast.* This last association encapsulates the theme of this chapter: that much of our early experience is shaped by the universal infantile need for warmth. It is this need for warmth that has driven not only our physiological and behavioral responses but our language as well.

6 FEVER ALL THROUGH THE NIGHT

THE WORLDWIDE USE of flu vaccines has resulted from a concerted effort among international health workers to combat the novel, virulent strains of the influenza virus that pop up each year. I will never forget the day in 1989 when I was infected with one such strain—the Shanghai. One moment I was sitting on my couch listening to music, and the next I had slithered to the floor like a rag doll, where I lay, breathing deeply, moaning, barely able to speak. At the time, I did not know what had hit me, but I eventually came to understand that all of my bodily effort and energy had been focused on one goal: producing a fever.

Whether prompted by a rare strain of the influenza virus or a bacterial infection caused by eating bad meat, fevers seem, to many of us, to be part of the problem. Indeed, we often think of the viral or bacterial agent as *causing* the fever. But that's not an accurate assessment of the situation. The fever is actually a defensive weapon against the invading virus or bacterium. Fevers are an ancient bodily response to infection, part of the survival arsenal of animals as varied as humans, iguanas, and lobsters. Despite this ancient history, our appreciation of the substantial benefits of fever is only a few decades old.

Both fever and hyperthermia entail an increase in body temperature, but the similarities end there. If you have ever experienced a fever, you may recall suddenly feeling cold, shivering, curling up in a ball, retreating under the covers of your bed, and drinking hot liquids. As noted earlier, these are *heat-gain* responses: your body is trying to warm up. In contrast, when your body temperature increases because of intense exercise or sunbathing, you feel hot, not cold, prompting you to exhibit *heat-loss* responses such as sweating, sprawling, drinking cool liquids, and jumping into a pool. The former situation is referred to as a fever, the latter as hyperthermia.

We distinguish between these two situations—fever and hyperthermia—on the basis of the correspondence between the change in body temperature and the animal's thermoregulatory responses. In the case of fever, heat-gain responses work to promote the increased body temperature, while in the case of hyperthermia, heat-loss responses work to counteract the increased body temperature.

We can use this same scheme to distinguish between different causes of decreased body temperature. One cause is the opposite of hyperthermia: hypothermia. For example, a few years ago my dog Aja began exhibiting some strange symptoms. Without warning, she would have fits of uncontrollable shaking, especially of the head. At first it looked like epilepsy, but when I placed my hand on the large muscle that holds up her head, I noticed that it was very warm. Aja was apparently shivering. If I warmed her up (she loves the hot tub), the shivering stopped. It turned out that her thyroid gland had stopped producing sufficient amounts of the hormone thyroxine; this condition is known as hypothyroidism and is treated with hormone replacement pills. Thyroxine has a number of functions, one of which is to stimulate the utilization of energy (and therefore the production of heat) in every cell in the body; in a sense, this hormone establishes the body's minimum rate of heat production. When her thyroxine levels dropped, Aja's body and brain temperatures fell, whereupon her hypothalamus triggered shivering as a means of increasing body temperature back to normal levels. In other words, by shivering, Aja was attempting to counteract the decrease in body temperature. This is hypothermia.

Just as hyperthermia has a counterpart in fever, so hypothermia has a counterpart. We don't have a word for it, so let's call it anti-fever. As an example of anti-fever, consider the hot flash (or flush), a disturbance of the thermoregulatory system that is experienced, to varying degrees, by 75 percent of women in the early stages of menopause. For reasons that are not fully understood, the change in hormonal status that accompanies menopause can occasionally trigger a sudden intense feeling of warmth, eventuating in a burning sensation in the face, neck, and chest brought on by rapid dilation of the blood vessels in the skin in those areas. Women respond to these flashes by engaging in a variety of heat-loss behaviors, such as removing clothing, throwing off blankets, opening a window, or standing near a fan or air conditioner. In this case, a feeling of warmth activates heat-loss responses that work to decrease body temperature.

To further examine these relations between body temperature, feelings of warmth and cold, and thermoregulatory responding, consider the common household thermostat. A thermostat consists of a temperature-sensing device and a dial. The dial is used to establish the *set-point temperature* of the room, say 22°C (72°F). The thermostat is connected to a switch that determines whether the furnace in your house will be activated and hot air will be circulated throughout the house. If the house temperature measured at the thermostat is lower than 22°C, then the compressor will be activated until the house temperature has been driven up to the set-point temperature of 22°C, at which point the compressor will shut off. Now, if you decide that you want the house to be hotter, you will turn the dial on the thermostat to a higher set-point, say 25°C (77°F), and the compressor will kick on and warm air will be circulated until the new set-point is reached, after which the compressor will turn off and on as needed to maintain a room temperature that conforms to this new set-point temperature.

(The concept of set-point, although useful as a metaphor for understanding some aspects of thermoregulation in animals, should not be confused with the actual neural mechanisms by which we and other animals accomplish thermoregulation. Those mechanisms are still largely unknown.)

We can use the set-point metaphor to illustrate the entire fever process. Viral or bacterial infection results in an increase in your set-point temperature and the initiation of the *chill phase* of fever. During this phase, the set-point temperature rises to a level above body temperature, a situation that is associated with feeling cold and with behavioral and physiological heat-gain responses, such as shivering. In time, these responses successfully increase body temperature to the febrile (feverish) set-point temperature, at which point you feel normal, at least from a thermoregulatory standpoint. Perhaps, in the middle of the night, you wake up drenched in sweat and throw off the covers: this indicates that the fever has broken, or, in other words, that the body's set-point temperature has suddenly returned to its normal level, but the body's actual temperature is still elevated. Feelings of warmth and heat-loss responses are triggered when set-point temperature is lower than body temperature. When these heat-loss responses have eliminated the differential between the two temperatures, the cycle is complete and regulation of body temperature resumes at the normal set-point level.

We can also use the concept of set-point to illustrate another feature of fever. When a person or another animal is infected and produces a fever, the maximum body temperature generated during the fever is unaffected by the surrounding air temperature. What vary with external temperature are the specific heat-gain mechanisms that are used to generate the fever. For example, if a monkey is housed in a warm environment, it will dilate its peripheral blood vessels to increase heat loss and maintain normal body temperature; if it is then infected, it will constrict its peripheral blood vessels, thereby decreasing heat loss through the skin and generating a fever through heat retention. In contrast, if the monkey is housed in a cold environment, it will constrict its peripheral blood vessels to maintain normal body temperature; consequently, when it becomes infected, further constriction is not possible.

Under these circumstances the monkey will generate a fever by increasing its internal heat production (perhaps by shivering). Furthermore, if you were to give a different monkey, housed in a cold environment, a string that it could pull to turn on a heat lamp, the monkey would learn to pull the string to increase its body temperature, a more energy-efficient way to produce a fever than by activating energy-intensive mechanisms like shivering. Again, infection appears to result in a raised set-point temperature that in turn evokes any of a number of physiological and behavioral responses to generate a fever. The particular responses recruited depend upon the environmental context and the behavioral choices available to the animal.

INFECTION, FEVER, AND SURVIVAL

That fever is an ancient response to viral or bacterial infection suggests, but does not demonstrate, that it is beneficial. Remarkably, convincing evidence that fever is beneficial did not begin to accumulate until the mid-1970s, when the physiologist Matthew Kluger began his investigations of the contributions of fever to survival. For their groundbreaking work, Kluger and his colleagues studied the desert iguana, an ectotherm. As it turned out, studying a feverish lizard yielded new knowledge about the function of fever in mammals, including humans.

It was, in fact, Kluger's decision to study an ectothermic lizard rather than an endothermic mammal that led to the key initial insight into the function of fever. Because ectotherms cannot, by definition, increase heat production internally, investigators can easily manipulate their level of fever by manipulating the temperature of their environment. In contrast, it is very difficult to manipulate fever levels in mammals without using pharmacological agents, such as aspirin, that may have other effects on the animal's response to infection, thereby making it difficult to draw sound conclusions regarding the function of fever itself.

The first step in Kluger's experimental approach was to establish the level of fever exhibited by desert iguanas infected with a bacterium. Iguanas were placed in an environment in which they could select, throughout the day and night, from a wide range of environmental temperatures while their body temperature was monitored; average body temperature during this first, pre-infection day was 38°C (100°F). On the second day the iguanas were injected with dead bacteria of the species *Aeromonas hydrophila* (the injection of dead bacteria triggers a febrile response but does not threaten the animal's health). After being infected, the iguanas were allowed to continue thermoregulating behaviorally. Within five hours, they began spending more time in the warmer regions of the apparatus, and their body temperature quickly rose to 42°C

(108°F), where it remained throughout the third day of testing. This experiment demonstrated that, when given a choice, infected iguanas will produce and maintain febrile temperatures that are 4°C (7°F) above their normal preferred body temperature.

Kluger's next experiment is a classic. He injected iguanas with fever-producing bacteria (this time the bacteria were alive) and then placed the iguanas in homogeneous thermal environments that did not allow for behavioral thermoregulation. Some were housed at a temperature of 42°C (108°F), the febrile temperature iguanas had selected naturally in the earlier experiment, while others were housed at various temperatures from 34°C (93°F) to 40°C (104°F). Within three days after being infected, nearly all of the iguanas maintained at 34°C died, whereas nearly all of those maintained at 42°C (108°F) survived. Iguanas at the intermediate temperatures exhibited intermediate rates of survival. This experiment provided the first convincing demonstration that fever can be beneficial to the host and is not merely a symptom of disease.

When we get a fever, we typically reach for an *antipyretic,* that is, a drug such as aspirin, ibuprofen (for example, Advil), or acetaminophen (for example, Tylenol) that returns body temperature to normal levels. The underlying principle, we have been taught to believe, is that fevers are undesirable and should be treated, just as we might treat the headache, the runny nose, and the cough that accompany a cold. This negative opinion of fever is relatively recent: many ancient writers, including Hippocrates, considered fever a beneficial bodily response. This view held sway for many centuries, but over the past few hundred years fever lost its good reputation. Our current, almost reflexive use of antipyretics reflects the widely held view that fever does not benefit the patient.

At best, the belief that fever is a mere symptom of illness is misleading—and it may be dangerous. One study performed by Kluger and his colleagues is sufficient to tell the tale. Twelve desert iguanas, housed in an environment in which a wide variety of temperatures was available, were infected and simultaneously treated with sodium salicylate, an aspirin-like antipyretic agent. In spite of the antipyretic, five of the iguanas developed fevers, and all five survived. In contrast, all seven iguanas that did not develop fevers succumbed to the infection and died. When an additional eight iguanas were infected, injected with sodium salicylate, and placed in a constant fever-causing environment maintained at 41°C (106°F), only one died. This experiment demonstrated clearly that, in iguanas, administration of an antipyretic can be harmful, *especially* when it succeeds in reducing a fever.

But is aspirin also harmful when given to infected mammals such as rats, rabbits, and humans? For these animals the story is more complicated. When the febrile responses of infected rabbits were measured and related to their rates of survival, it was found that a rabbit's chance of surviving infection was 60 percent for moderate fevers of 1.50–2.25°C (2.7–4.0°F), whereas the chance of survival dropped to only 20 percent for higher or lower fevers. This experiment showed that, in the right amount, fever is a good thing for rabbits.

The results of another experiment cloud this picture a bit. Investigators again infected rabbits with a bacterium and then treated them with antipyretics. Although fevers greater than 2.25°C were not observed in this experiment, once again rates of survival increased when moderate fevers were produced. There was also an added boost in survival rates in the animals given antipyretics that could not be attributed to the fever, suggesting that the drugs had benefits beyond their action as antipyretics. Nonetheless, regardless of whether or not a rabbit was treated with antipyretics, a fever in the range of 1.50–2.25°C was associated with the highest rates of survival.

These experiments demonstrating the benefits of fever were conducted over twenty years ago and have yet to be disputed; on the contrary, they have been supported repeatedly by studies using ectotherms and endotherms alike, including studies of humans. Despite these findings, we continue to perceive fever as a symptom of disease rather than as an ally in the fight against disease. Clearly this attitude is no longer warranted, although our understanding of the benefits of fever in humans and the usefulness of aspirin and similar drugs remains incomplete.

THE FEVER CASCADE

As we have seen, fever is a well-orchestrated thermoregulatory process by which an animal uses physiological and/or behavioral mechanisms to raise its body temperature. But fever is also part of a larger cascade of processes—the acute phase response—that in its entirety appears designed to fight infection.

Infection begins when a pathogen (any microorganism or virus that causes disease; from the Greek *pathos,* meaning suffering, and *genes,* meaning producing) pierces the protective barriers that stand between us and the outside world—that is, the skin as well as the linings of the respiratory and digestive systems. Once the pathogen has gained access to the circulatory system, the body's immune system is engaged: cells called leukocytes (or white blood cells) and neutrophils move swiftly to the site of the infection, where they engulf the pathogenic cells and begin to destroy them. In the midst of this activity, the invading immune cells release a protein molecule that initiates the acute phase response. This molecule, called interleukin-1 (IL-1), has many known effects, including the generation of fever, the suppression of food intake (a response that is consistent with the admonition to "feed a cold; starve a fever"), and the activation of the immune cells that help us fight viral and bacterial infections. IL-1 has also been shown to increase sleep, an effect that appears to account for the drowsiness we commonly experience when we are sick.

The fever cascade continues as IL-1 travels through the circulatory system to the brain. At this point it encounters a barrier.

The brain is a privileged organ in many ways. It is protected physically by the skull and a tough outer lining, and it is protected in terms of energy by having primary access to glucose. The brain is also protected chemically by a barrier—called the blood-brain barrier—that prevents many molecules in the circulatory system from gaining entry. Investigators first discovered this barrier by injecting blue ink into an animal's circulatory system; upon autopsy, it was found that every organ had turned blue except the brain.

Despite this barrier, the brain does monitor the circulatory system, and one way it does this is with a group of organs called circumventricular organs, which, like antennae, poke through the protective blood-brain barrier with specialized detectors and cells. According to one model of IL-1's action in the generation of fever, IL-1 enters a circumventricular organ, where it induces cells to produce PGE_1, an agent that belongs to a class of molecules called prostaglandins. Then PGE_1 either diffuses into a nearby brain region or acts on neurons within the circumventricular organ to continue the fever cascade.

Regardless, the cascade is completed when neural activity within the hypothalamus is altered and the nervous system produces the behavioral and/or physiological heat-gain responses that are ultimately responsible for the generation of fever.

This rough outline of the many molecules and processes that participate in the fever response is sufficient to drive home the fundamental point that began this chapter: fevers are produced by the host, not by the pathogen. The host synthesizes endogenous pyrogens (from the Greek *pyros,* meaning fire, and *genes,* meaning producing)—molecules such as IL-1 and PGE_1—and organizes neural and physiological systems so that febrile temperatures can be produced, maintained, and defended. These endogenous pyrogens are ancient molecules. PGE_1 is present in such distant evolutionary relations as scorpions, shrimp, and crabs, and when injected into these animals it will produce a fever, just as it does when infused into the brain of a rat or a rabbit. The commitment of an organism's energy to producing a fever, the findings of fever's benefits to the host, and fever's ubiquity among diverse organisms have convinced many scientists of its functional importance.

But how does fever promote survival? If fever is such an ancient response to infection, one would think that the mechanism by which it benefits the host would be easy to determine. In fact, it has been difficult, in part because temperature influences many aspects of immune function, both directly and indirectly.

For example, recall that the body's initial response to infection is the assault by specialized cells that engulf and destroy foreign bodies such as bacteria and viruses. This immune response is aided by the production and secretion of a variety of antibacterial agents, including hydrogen peroxide. As it turns out, increases in body temperature typical of those produced during fever result in accelerated mobilization of immune cells to the site of infection as well as in swifter secretion of antibacterial agents.

Recall also that the initial attack by leukocytes and neutrophils triggers the secretion of IL-1, the endogenous pyrogen that has a variety of beneficial effects, including the activation of the immune cells, including T cells, that fight infection. This enhancement of immune function is a direct effect of IL-1. But the fever triggered by IL-1 also has an indirect effect on this process by stimulating the production of more T cells.

IL-1 secretion has yet another function—the reduction of iron levels in the blood. Iron is a vital mineral for humans and other animals with circulatory systems because of its role in binding oxygen to hemoglobin so that oxygenated blood can be delivered to tissues throughout the body. But iron is also an essential nutrient for many species of bacteria, and the dependence of these bacteria on iron increases at febrile temperatures. Therefore IL-1, by reducing iron levels in the blood of the host, deprives the invader of an essential nutrient.

It is evident that fever helps an animal fight infection. But if fever is so beneficial, why haven't we endotherms—mammals and birds—evolved to regulate body temperature at febrile temperatures and thereby decrease the ability of pathogens to invade and infect us? Although the answer to this question is not known with certainty, some probable reasons are apparent.

First, we endotherms already devote a great deal of our energy budget to the regulation of body temperature; if we were to regulate body temperature 2°C (4°F) higher than we do now, our energy consumption would rise by approximately 20 percent, a sizable increase that would necessitate significant adjustments in our eating and sleeping habits. Second, constant regulation at higher temperatures would bring us even closer to those temperatures, 41°C (106°F) and above, at which the integrity of body and brain becomes difficult to sustain, and thus would make it even more crucial for animals to ensure protection against overheating. Third, because increased temperatures during pregnancy lead to serious developmental disorders in offspring, substantial changes in gene expression during development would be required. And finally, the chronic stimulation of the immune system that would result from a higher regulated body temperature might increase animals' susceptibility to autoimmune diseases, such as multiple sclerosis and lupus, in which the immune system becomes overactive and attacks body tissue as if it were a foreign invader. For these and other reasons, evolving even a small increase in body temperature would have costs as well as benefits for an organism.

Every species regulates body temperature, behaviorally or physiologically, within a range that balances the needs of a wide variety of systems, all of which are temperature-sensitive. Any change in such a fundamental variable as temperature will have far-reaching consequences; some of them will be beneficial, some detrimental. For hundreds of millions of years, animals have benefited from the evolutionary discovery that, regardless of the normal body temperature of a species, just a little rise in temperature for a short time increases an individual's odds of surviving an onslaught by a viral or bacterial pathogen.

7

THE HEAT OF PASSION

I HAVE A FRIEND WHO has a passion for peppers. He grows many varieties, including jalapeños, serranos, and cayennes. A few summers ago he hosted a barbecue, and the conversation inevitably turned to that year's crop. In due time, a gustatory game of competitive pepper-eating ensued, and for this game the big guns were brought out—the habaneros. One by one, each brave warrior took a bite. The pain became obvious in short order, but the macho veterans of this game refused to give in, even as their faces glowed, as sweat dripped down their faces, and as water leaked from their eyes. The novices, however, surrendered quickly, blurting out something like, "Wow, that's a hot one!"

Why do we say that hot peppers are *hot?* Judging from my own experiences with peppers, it does seem as if I have eaten something hot: I may open my mouth and wave my hand in front of it, and I may quickly drink a glass of water while exclaiming that my mouth is on fire. My mouth *feels* hot, although neither hand-waving nor cool water is very effective at putting out this fire. In addition to the burning sensation in my mouth, responses throughout my body also suggest a thermal experience. I feel overheated. Most notably, my face flushes and I begin to sweat. What's going on here?

SPICY HOT OR TEMPERATURE HOT?

All peppers are members of the nightshade family, along with tomatoes and potatoes, and have been cultivated throughout Asia as well as Central and South America for hundreds of years. What makes peppers taste hot is the presence of a compound, concentrated in the veins of the fruit, called *capsaicin*—the higher the concentration of capsaicin, the hotter the pepper. After many years of selective breeding, there are now a multitude of pepper varieties, ranging in "heat" from mild to fiery.

When we speak of hot peppers and hot potatoes, we mean different things. The potato's heat can be measured with a thermometer. The pepper's, however, cannot. Instead, thanks to Wilbur Scoville's invention of 1912, we measure the heat in chili peppers with the Scoville Organoleptic Test. The Scoville test is a classic example of psychophysical measurement: A pepper is ground up and diluted in a solution of sugar water and alcohol. Human subjects taste this solution and report whether or not they can detect the pepper's heat, and the amount of sugar and alcohol is gradually increased until three of every five tasters (or 60 percent) report that they cannot. When that 60 percent threshold is reached, the number of dilution units is noted, and this value, in Scoville Units (S.U.), indicates the heat in the pepper. The more dilution required to remove the heat, the higher the pepper's S.U. value. Thus bell peppers have a value of 0 S.U.; jalapeños, about 5,000 S.U.; serranos, as high as 25,000 S.U.; and habaneros, as high as 300,000 S.U. Pepper spray, used by police and civilians to aid in arrests or to ward off attackers, typically has a value of 2 million S.U. Pure capsaicin has a value of about 16 million S.U.

Just as my mother used the back of her hand applied to my forehead to determine if I had a fever, the Scoville Test uses the human tasting machinery to measure the concentration of capsaicin in a pepper. Clearly, neither approach is as accurate or objective as a thermometer or a liquid chromatograph for measuring the concentration of heat or capsaicin. The value of the Scoville Test, however, is precisely its subjectivity—it assesses the sensitivity of the human tasting apparatus and, in doing so, reminds us that a pepper's concentration of capsaicin alone does not predict how an individual will respond to the pepper. Indeed, there is enormous variability in human tolerance for capsaicin: some people need only taste a habanero once to know never to go near it again, while others pop habaneros into their mouths with little concern.

The effect that capsaicin has on receptors in the mouth is part of a much broader phenomenon. Capsaicin is irritating when it comes into contact with any of a number of places on our bodies. Those who carry pepper spray as a defense are instructed to aim at an assailant's face, especially the eyes. Also, experienced gardeners and cooks wear gloves when handling the hottest pepper varieties because an open cut can burn terribly when it comes into contact with a pepper. It appears, then, that many parts of our body are sensitive to capsaicin, not just our mouths.

Our bodies are equipped with detectors that sense stimuli in the outside world and then convey sensory information via nerves to our brain. Detectors in our eyes are sensitive to light; detectors in our ears are sensitive to sound; detectors in our nose and on our tongue are sensitive to chemicals conveyed in the air, suffused in food, and dissolved in drink. We also have various kinds of detectors in our skin that are sensitive to touch, vibration, and, of course, temperature. Normally these detectors are activated only by the particular stimuli that they were designed, through the evolutionary process, to detect. But we also know, for example, that artificial electrical stimulation of a light detector in the eye will evoke the sensation of light *even in the absence of light.* The brain bases its perceptions of the external world upon information provided by these detectors, regardless of whether they have been activated by a natural stimulus or by a scientist with an electrode.

Noxious or painful stimuli are transmitted to the brain by *nociceptors,* specialized detectors located in the skin. Among other things, nociceptors are sensitive to intense heat, such as that experienced when we touch a hot stove. One of the critical sensory events that trigger the communication of this information about noxious heat to the central nervous system depends upon a molecular mechanism that is sensitive to capsaicin. In other words, capsaicin gives us the feeling of a burning mouth, burning eyes, or burning skin because it is activating sensory nerves that are used by our bodies to transmit information about intense thermal stimulation. We say peppers are hot because they evoke the same sensations as intense heat.

Capsaicin does more than activate sensory nerve endings. When we eat a pepper, capsaicin makes its way into our bloodstream and eventually into our brain, where it can have effects lasting for hours. One effect of injecting capsaicin into an adult rat is dilation of the peripheral blood vessels, which probably explains the facial flushing that many of us experience when eating hot peppers. When these vessels dilate, our skin feels hot because there is increased heat flow through our skin surface, which activates the temperature detectors there.

Moreover, as a result of this increased heat flow, capsaicin also causes a prolonged decrease in body temperature. (This feeling of being warm even as body temperature falls may be similar to the effect of overconsumption of alcohol, another potent dilator of peripheral blood vessels; this effect can be deadly for alcoholics living on the street on cold nights, as they may feel thermally comfortable even as their body temperature plummets to lethal levels.)

Capsaicin's effects are mediated, at least in part, by the brain. For example, injecting capsaicin directly into the brain of a rat alters the activity of neurons within a part of the hypothalamus that plays a role in organizing the body's response to overheating, thereby triggering peripheral vasodilation, including facial flushing. (These responses are similar to a hot flash, although no one to my knowledge has directly compared the two phenomena.)

Investigators have recently made tremendous strides in understanding the role of capsaicin-sensitive neurons in the nervous system by developing a genetically modified mouse that lacks the ability to respond to capsaicin. Not only are these mice insensitive to the taste of the hottest peppers, they are also impaired in their ability to detect intense thermal heat that normal mice perceive as painful. Thus, even though hot peppers are not literally hot—high in temperature—on our tongues, our language long ago came to reflect a physiological reality whose basis has been revealed by scientists only in the last two decades.

HEATED WORDS

That we have come to refer to spicy foods as hot testifies to the flexibility of our language for reflecting our physiological responses to food. But we also use thermal metaphors to refer to a variety of human passions. For example, we use such metaphors to describe a person's volatility or proclivity toward violence. Someone who has a tendency to lose his self-control, or has already lost it, is called a *hothead, hot tempered,* or *hot and bothered.* He may be *boiling with anger* or *steaming with rage* after someone has pushed his *hot button.* On the other side of the thermal spectrum, people who exhibit self-control under difficult conditions are *keeping a cool head, cool under fire,* or, in more contemporary parlance, *chillin'.*

The connection between temperature and violence may be more than mere metaphor. The psychologist Craig Anderson has argued, albeit using mostly correlational evidence, that aggression and violence increase under high-heat conditions, independent of geographical location or socioeconomic status. Evidence in support of this notion, cited by Anderson, includes the greater tendency of baseball pitchers to hit a batter with the ball on hot days and increased rates of horn honking on hot days by drivers whose cars lack air conditioning.

Anderson has even suggested that, if current global warming trends continue, an increase in average temperature by 1°C (2°F) will result in 24,000 additional murders each year in the United States.

The idea of a connection between climate and violence is not new: in *The Spirit of Laws,* published in 1748, the French philosopher Montesquieu noted that for the inhabitants of hot, southerly climates, "the strongest passions multiply all manner of crimes." Montesquieu wrote these words as part of a broader treatment of the effect of climate on human behavior, a treatment that included, of course, sex: "In northern climates scarce has the animal part of love a power of making itself felt . . . In warmer climates love is liked for its own sake, it is the only cause of happiness, it is life itself." Montesquieu even proposed a biological mechanism connecting climate with passion and pleasure: "In cold countries the nervous glands . . . sink deeper into their sheaths, or they are sheltered from the action of external objects." In other words, the cold causes the sensory nerve endings, like moles hiding in their burrows on a chilly day, to stay inside where it's warm. As a consequence, even though Montesquieu thought the inhabitants of cold climates to be more vigorous, noble, and virtuous, he argued that "they have very little sensibility for pleasures."

Despite the simplicity of Montesquieu's climatic and physiological analyses, he was correct in recognizing a fundamental relationship between temperature and sex. Indeed, our language reflects this relationship through the ubiquity of sexual-thermal metaphors. “I’ve got the hots for you,” “hot pants,” “I’m burning with love,” and “she’s frigid” are just a few examples. The connection between sex and temperature has often attracted the attention of songwriters, as with Cole Porter’s lyric “This love affair was too hot not to cool down” or Peggy Lee singing “He gives me fever, fever all through the night.” The Pointer Sisters voiced a similar sentiment with the lyric “When we kissed—fire.” The rock band Foreigner scored two thermal hits with “Cold as Ice” and “Hot Blooded” and Donna Summer had one of her own with “Hot Stuff.” And of course there is Mick Jagger shrieking, “I’m so hot for her, I’m so hot for her, I’m so hot for her, and she’s so cold.”

As we saw with peppers, our language can provide an accurate reflection of underlying physiological processes. Might the pervasive use of sexual-thermal metaphors also reflect a connection between sex and thermal physiology? The answer to this question is not as obvious as one might think, because there are a variety of ways in which thermal metaphors are used. For example, as mentioned earlier, thermal metaphors can reflect an assessment of another person’s thermal state (*hot-headed*). Alternatively, these metaphors can be used to describe the openness of another person to physical or emotional contact (*warm, cold*).

We also use thermal metaphors in a third way: to describe our physiological reactions to stimuli in the environment—inanimate as well as animate. On the television show *Who Wants to Be a Millionaire?* contestants are invited to sit in the *hot seat,* a term that reveals the increased heat production, accompanied by sweating, that accompanies the stress of the game. Of course, the seat itself is no more literally hot than a pepper is, but this does not prevent us from projecting the heat onto the seat. Do we do the same with people? That is, are people we find sexually *hot* actually triggering physiological reactions that make *us* feel hot? And are these physiological reactions during or in anticipation of sex the basis for the creation and retention of sexual-thermal metaphors in our language?

HOT FOR YOU

Oddly enough, we have very little scientific information about the thermal changes that accompany sexual activity in humans. We don't need science, however, to tell us how we feel during sex. As with other strenuous activities, we heat up during sex as a result of intense muscle activity. We also breathe more intensely, just as we do during exercise, and we may sweat.

What may be less obvious to us is the supreme importance of blood flow to sexual behavior—specifically, the importance of directing blood flow to the genitalia. Consider that the primary effect of Viagra, perhaps the most popular drug among men on the planet, is to increase the flow of blood to the penis. Increased genital blood flow provides the basis for penile engorgement and erection, as well as heightened sensitivity of the genital regions of both men and women for the achievement of orgasm. Because there is a fixed volume of blood in the body at any given time, the extra blood flow to the genitalia must come at the expense of blood flow to other areas. This redistribution can be achieved through focused constriction and dilation of blood vessels throughout the body. The intense muscle contractions in the extremities, especially the arms and legs, during sex also help to move blood. Moreover, blood flow and oxygen supply to the brain probably decrease during sex, a phenomenon that may generate feelings resembling the euphoria that is known to accompany the decreased availability of oxygen to the brain associated with hypoxia. (Some people attempt to enhance the pleasure of sex by further reducing blood flow to the brain using a rope tied around the neck, sometimes with lethal consequences.) The reduced blood flow to the brain helps to explain why some people feel dizzy after an orgasm, especially if they rise quickly from a horizontal position.

Thus far, the best evidence concerning the thermal changes occurring during sex comes from studies on rats. To many humans, rat sex may seem a bizarre concept, an odd combination of what is perhaps the world's most detested animal with the ultimate pleasurable activity. To biopsychologists, however, rat sex is both elegant and complex, a Darwinian dance that has provided the foundation for some of the most important advances in our understanding of the hormonal and neural mechanisms that guide and shape animal behavior, including our own.

Scientists study the rat's sex life by sitting in a dark room in the middle of the night and watching the animals interact in a small box illuminated with red light. Red lighting is used neither for ambiance nor to pay homage to Amsterdam's renowned district of ill repute: rats are nocturnal and are not disturbed by red light, thus allowing us to observe their behavior.

When a male gains access to a female, the two rats investigate each other, largely through sniffing. The female engages in *proceptive* behaviors in which she solicits the male's attention. She will nose and nudge him and run away when approached. In time these behaviors stimulate the male to chase after the female, while she hops and darts, staying just in front of the increasingly active male. At some point, with the male in pursuit, she stops abruptly and braces herself, and the male catches her and mounts her from behind. As he mounts her, he grabs her flanks with his forepaws, a stimulus that triggers the female to arch her back in a way that allows penile insertion (this reflex can be activated only when she is in the right hormonal state). If the male achieves penile insertion, or *intromission,* he does not ejaculate but rather quickly lunges backward and grooms himself.

Once the first intromission is achieved, the pace of the behavior accelerates. She solicits, and he chases, intromits, and grooms himself. This pattern continues over the next few minutes, until ejaculation is imminent. He lumbers toward her, she continues her hopping and darting, he mounts lethargically, and with a kick of his hind leg he ejaculates, whereupon he releases his grasp of the female, and she scurries away.

After the male ejaculates, he will often stroll to the corner of the cage, sprawl on the floor, and begin to emit an ultrasonic vocalization (the communicatory significance of this vocal behavior, if it has any, is not yet well understood). After a few minutes of this post-ejaculatory interval, both male and female are ready to resume. For this second go-round, the pattern of behavior is similar, but now the male requires less time to ejaculate and more time to recover. And so on, and so on, for several hours, until both male and female have had enough.

Male rats heat up during sex. In one study, my colleagues and I found that as soon as the female was introduced into the male's cage, his body and brain temperatures began to increase. By the time the male ejaculated, his temperature had risen more than 1°C (2°F). But then, to our surprise, we found that his brain, but not necessarily his body, cooled down following ejaculation, only to heat up again when copulation resumed. This pattern of increasing and decreasing brain temperature during sex turned out to be a highly reliable finding.

Why do male rats heat up during sex? One possible answer is that sex entails vigorous muscle activity, so maybe sex is just a form of exercise. We found, however, that mere exercise does not cause a male rat to heat up nearly as quickly or as easily as sex does. It appeared, then, that exercise could not account for all of the thermal changes that accompany sex. Rather, as we soon discovered, rats exhibit massive constriction of the peripheral blood vessels in the head (and probably elsewhere) immediately upon exposure to a female, before physical contact is made and before the male's activity increases. Such constriction reduces the flow of cool blood from the skin surface of the head and face to the brain, thereby causing a rise in brain temperature. This constriction is maintained until ejaculation, after which the blood vessels dilate quickly and the brain cools. Then, when copulation resumes, vessel constriction and brain heating recur.

The constriction of the rat's blood vessels during sex is probably a part of the general strategy to direct blood flow to the genitalia. If so, then the heating of the brain during sex may simply be a by-product of this need to send blood to the (temporarily) more important organ. This need vanishes upon ejaculation, at which time normal blood flow to the brain can resume and the brain can cool.

Female rats also heat up during sex, although they do not exhibit rapid brain cooling when the male ejaculates, nor do they sprawl or exhibit other signs of being overheated. Indeed, despite all their exuberant hopping and darting, female rats seem to be far less physically challenged during sex and less exhausted by the end of it than males. The reason for this is not clear.

Peggy Lee notwithstanding, that a rat gets hot during sex does not necessarily mean it has a fever—sex may simply entail hyperthermia. In fact, there is reason to believe that sex does not produce a true fever. Recall that a male rat sprawls after ejaculation when his peripheral blood vessels dilate and his brain cools. Sprawling is a heat-loss behavior, a sign that the rat *feels* hot. This sounds like a hyperthermic rat, not a feverish one. And yet sprawling occurs *after* ejaculation—so perhaps the male rat gets a fever during sex which turns into hyperthermia after ejaculation. We simply don't know.

LONG-DISTANCE LOVE

Although it seems like a good bet that our language has come to designate hot peppers and hot people for similar physiological reasons, eating peppers and engaging in sex differ, and not just in the obvious ways. Perhaps most important, consumption of peppers has not played an important role in human evolution. In contrast, sex (along with other behaviors such as drinking and eating—of carbohydrates and fat, not capsaicin) is one of the so-called motivated behaviors that play vital roles in animals' survival. Motivated behaviors such as sex are highly complex, but they are built upon a foundation of simple reflexes; through Pavlovian conditioning over the course of a lifetime, these reflexes can come to be triggered by a variety of sights, sounds, or smells that were once sexually neutral but that we have learned to associate with sexual activity. In contrast, we don't get hot when we merely see a pepper (although we do learn the association between certain pepper shapes and colors and their degree of heat).

To appreciate this, let's take a closer look at eating—not peppers, but candy. When we eat candy, our bodies experience a number of changes that help us to digest it. Our mouths fill with saliva to help break down the glucose in the candy, and our stomachs release gastric acid to aid digestion. In addition, the pancreas secretes insulin into the blood. Insulin is a hormone that serves two functions: it makes it possible for cells throughout the body to utilize the sugar in the candy as energy, and it converts excess sugar that is not needed immediately for energy into glycogen (starch) and, eventually, into fat. Because insulin moves glucose out of the bloodstream, and because there are sugar detectors (for example, in the liver and brain) that help the body monitor and regulate the amount of glucose available as a source of energy, insulin is capable, when injected into a normal person, of inducing a state of hunger.

When we are young and have had no experience with candy, the candy must be placed in our mouth to stimulate these salivary, gastric, and pancreatic responses. But as we grow older and gain experience with candy, these physiological responses come to be elicited before the actual act of eating. As with Pavlov's dogs, these responses become conditioned to the mere sight of the candy wrapper or the smell of the candy or the mention of its name. Candy distributors take advantage of this conditioned response by building candy machines with glass windows that prominently display their contents. Similarly, proprietors of movie theaters use advertisements to encourage the moviegoer to imagine the sight, smell, and taste of hot buttered popcorn, providing images and sounds in case the imagination fails.

What purpose does this sensory bombardment serve? Because of our lifetime of acquiring conditioned responses to food, we respond physiologically to the sensory predictors of food before we actually put the food in our mouth; these responses are particularly strong when it has been a long time since we last ate. Such conditioning makes perfect sense from an evolutionary point of view, because activating these physiological responses before eating aids in the digestion of the soon-to-be-eaten candy or popcorn. But there is an added dimension to all of this—a self-fulfilling prophecy of sorts—in that these advance physiological responses also increase the likelihood of our purchasing the candy or popcorn. The secretion of saliva, the grumbling of the stomach, and the secretion of insulin into the blood all intensify the drive to eat. Insulin's role is particularly important because of its ability to remove glucose from the blood and thereby stimulate the glucose detectors which trigger the brain state that we call hunger. The release of insulin exacerbates our perceived need for food and compels us to reach into our pocket, pull out the change, and purchase that luscious chocolate-and-caramel-and-nut-filled bar in that attractive red and white wrapper.

These examples illustrate that our physiological responses to food, especially food that contains an essential nutrient like sugar, can be activated at a distance as a result of a lifetime of learning. The same is undoubtedly true of our sexual responses: just as eating begins developmentally with the placement of food in the mouth of an infant and the consequent reflexive salivary, pancreatic, and gastric responses, so does sexual activity begin developmentally with genital rubbing and the consequent reflexive physiological changes that heighten arousal and pleasure. Thus, when we are young, the heat associated with sexual activity only arrives with the activity itself, and usually during solo explorations. But as we grow older and gain experience, our sexual and thermal responses can be triggered by any of a number of stimuli—sights, sounds, or smells. Simultaneously, we adopt the sexual-thermal metaphors prevalent in our language, usually without explicitly recognizing their physiological basis.

COMING FULL CIRCLE

We have now seen that peppers and people can both be designated as *hot* on the basis of the thermal effects that result from eating or from intimate sexual contact. Despite these similarities, however, peppers and sex do not mix. Recently a friend told me about a romantically involved couple she had known. (The couple, a man and a woman, are no longer together.) One evening the man cooked his sweetheart an elaborate Mexican dinner which contained a variety of peppers. Later, after dinner, the couple began to make love. He moved his pepper-contaminated hands across her body and, unfortunately, he ultimately reached her most sensitive zone. Much to the two lovers' dismay, they quickly discovered that he was too hot to handle her.

8 LIVIN' OFF THE FAT

My childhood physical exams with Dr. Silverman followed an established routine: I took off my shoes and stepped onto a shaky platform that formed the base of a hulking machine. First, to measure my height, the doctor extracted the long rod from its vertical sheath and extended the horizontal bar so it rested on the crown of my head. Then, to measure my weight, he adjusted the large, shiny pieces of metal and flicked the smallest weight with his index finger along the balance beam until the pointer on the far end of the bar floated in mid-air. With balance achieved, and with the information neatly recorded on a chart attached to his clipboard, the rest of the exam could begin, as if all subsequent medical decisions hinged upon this most basic of information, height and weight.

But although height and weight are similar in their simplicity, they are very different in how they rule our lives. Our adult height is fixed—there is little one can do about it, and consequently short people may suffer the occasional insult, but they rarely feel responsible for their "condition." Weight, however, is another matter. We are bombarded with the message that we can control our weight through willpower, proper nutrition, and exercise. The diet industry (which includes diet programs, diet foods, and diet magazines and books) is enormous, with revenues of tens of billions of dollars, catering to an American public desperate to slim down. We shop at grocery stores filled with food that isn't food and exercise at gyms packed with people walking furiously on treadmills that take them nowhere. Each morning we step onto the bathroom scale and measure our progress or lack thereof. And if we've failed, we feel miserable.

Our bodies are equipped with detectors and systems for maintaining a variety of physiological variables within normal limits. Blood pressure is monitored by a set of specialized receptors (called *baroreceptors*) that detect changes in the expansion and contraction of the arteries: if blood pressure increases and expansion is detected or if blood pressure decreases and contraction is detected, the baroreceptors trigger reflexive changes in heart rate to return blood pressure to normal. Similarly, the concentration of glucose in the blood is monitored by specialized receptors (called *glucoreceptors*) in the liver and the brain: when we eat a carbohydrate-rich meal, the glucoreceptors detect the increased blood sugar and trigger a change in nervous system activity that stimulates the pancreas to release insulin into the blood. Insulin then makes it possible for the glucose to be used as energy right away or to be stored as fat for later use.

Sometimes these systems break down. Arteriosclerosis, or hardening of the arteries, makes it difficult for the baroreceptors, and the cardiovascular system in general, to do their job of maintaining blood pressure within set limits, resulting in hypertension. And Type I, or early-onset, diabetes is a condition in which the pancreas has lost its ability to produce insulin, resulting in an inability to utilize glucose, which, if untreated, leads to soaring glucose levels and, ultimately, glucose poisoning.

Besides baroreceptors and glucoreceptors, our bodies rely on chemoreceptors for controlling the amount of carbon dioxide in the blood; osmoreceptors for controlling the amount of water available to our cells; and, as discussed in Chapter 4, thermoreceptors for controlling body temperature. When these systems fail, as in hypertension and diabetes, we purchase detectors from drug stores to help us accomplish consciously what is normally done without our awareness. If scales are analogous to these other purchased detectors, then their widespread use for monitoring body weight may seem to indicate an epidemic of failure in our body's system of weight detectors.

A little reflection will quickly dispel the notion that scales replace a function more properly performed by an internal weight-detection system. To measure your weight, you step onto a scale because a weight detector must be situated between the load-bearing part of your body and the center of the Earth. Therefore, if we possessed natural weight detectors in our bodies, they would have to be located in the soles of our feet or, perhaps more appropriately for some of us with desk jobs, in our buttocks. No such internal detectors exist. The conclusion is unavoidable: despite our fixation on body weight, our bodies couldn't care less.

Our bodies do, however, care about fat.

SEEKING BALANCE

When our distant ancestors were struggling for survival, food had to be hunted, gathered, and grown by all members of the social group. The availability of food was highly variable and unpredictable. When food was available, it was eaten, and what the body did not immediately use as fuel was stored as fat as a precaution against those inevitable periods when food was scarce.

But why store fat? When we eat more glucose than we can use, much of the excess is converted into an alternative form called glycogen and stored in the liver and in muscles as a short-term reserve. Why not simply store more glucose as glycogen as a long-term reserve? Or why not store protein? The reliance on fat for long-term storage derives from the simple fact that a pound of fat holds twice as much energy as does a pound of sugar or protein. Thus it is twice as energetically efficient to store fat than to store glucose or protein. Such efficiency can make the difference between life and death, especially when one must walk long distances between food sources.

The efficiency of fat as a fuel-storage mechanism is a general principle that applies to all animals, not just humans. This universal recruitment of fat for energy storage has not, however, required that the fat be deposited in the same place in all animals. Nor does it mean that fat serves only one function. Indeed, as we saw earlier, whales, seals, and other marine mammals have localized their fat as blubber under the skin to provide vital thermal insulation for survival in frigid waters.

Blubber is merely an amplification of the general pattern, found in nearly all terrestrial mammals and birds, of a layer of fat under the skin (subcutaneous fat) that serves as both energy reservoir and thermal insulation. For its role as an energy reservoir, the placement of fat is not critical. But for its role as thermal insulation, placement does matter. Small animals possess relatively thicker layers of subcutaneous fat than do larger animals. In large land mammals, such as cattle, fat is not concentrated subcutaneously, but is rather distributed throughout the muscle (referred to colloquially as marbling). An extreme evolutionary modification of fat localization is found in camels, whose humps are filled with fat (not with water); the thermal physiologist Knut Schmidt-Nielsen has noted that this concentration of fat in the hump may help camels to dissipate heat through the skin elsewhere on their bodies.

What we commonly call fat has two components: triglycerides (or lipids), the actual chemicals that provide the stored energy, and the fat cells that contain the triglycerides. The number of fat cells (that is, the containers) is established during development and remains stable throughout our lives; liposuction does not permanently remove these cells but only causes them to regenerate elsewhere. As we gain and lose weight, what we experience is the filling and emptying of the fat cells. But for most of us with ready access to food, such swings are rare and we regulate our amount of fat—and therefore our body weight—within a remarkably narrow range. In other words, like blood pressure and blood glucose, fat is a regulated resource.

To understand the system of fat regulation and the concept of energy balance, let's use a pool analogy again, as in Chapter 1. Imagine a pool filled to a depth of six feet. The pool has a drain and a faucet that, with the help of a device that measures the amount of water that flows through the drain, is adjusted to maintain the depth of the pool at exactly six feet. If we enlarge the drain, the faucet is adjusted to compensate. Similarly, if we open the faucet, a device instructs the drain to open up as well. Therefore, at all times, the input and output of water are kept in balance and the depth of the pool is regulated.

Similarly, to stay in energy balance, an animal must consume and utilize energy at the same rates. Energy flows into the system in the form of glucose, protein, and fat. Energy flows out of the system in a variety of ways: maintaining cellular metabolism uses energy, as do breathing, walking, thinking, talking, and even sleeping. To maintain energy balance, the body can adjust either side of the equation depending on the circumstances. If an animal overeats, the body can respond with increased physical activity. If an animal must raise its metabolic rate to stay warm in a cold environment, it can compensate by increasing its consumption of food. In the end, the amount of stored energy or fat, just like the depth of the pool, is tightly regulated. This response is the bane of dieters and explains why exercise is so important: if a dieter only decreases energy intake but does nothing to increase energy consumption (for example, by exercising), the body will compensate by decreasing energy consumption. Your body has evolved to protect its energy supply and doesn't much care how you look: mirrors are a relatively recent invention.

ENTER LEPTIN

But what is actually being regulated? Answering this question depended upon the successful breeding of rats and mice for obesity. Mice bred for obesity come in two widely studied varieties, referred to as *ob/ob* (for obese) and *db/db* (for diabetes); this nomenclature indicates that, as we have learned from breeding experiments akin to those developed by the Austrian monk Gregor Mendel in the nineteenth century, obesity is a recessive trait that is expressed when a mouse possesses two copies of the *ob* or *db* genes. Mice of both the *ob/ob* and *db/db* varieties exhibit threefold increases in body weight and fivefold increases in body fat (the animal on the left in Figure 8 is a *db/db* mouse), but the underlying cause for the obesity is not the same. Proving this required the use of a bizarre but ingenious experimental approach.

To appreciate the logic of this experimental approach, first imagine a signal that the body uses to inform the brain about its quantity of fat. The fatter you are, the more this fat signal is produced. To ensure that the brain responds to this signal and not to some other signal that serves another function, the brain contains specific receptors for this fat signal. Now imagine what would happen if, for some reason, the machinery that

Figure 8 A mouse bred for obesity *(left)* and a normal mouse *(right)*. The obese mouse lacks receptors in its brain for detecting leptin, the hormone that signals the quantity of fat present in the body.

produces the fat signal stopped working. Your brain would interpret this sudden loss of the signal as an indication that you were starved and would, in response, stimulate energy conservation and increased food consumption. But no matter how much you ate, the fat signal would not be produced and your brain would continue to function as if you were starved. Alternatively, imagine that the fat signal was produced normally but the receptors that detect it in the brain didn't exist or didn't function properly. Again, your brain would interpret your predicament as chronic starvation. Therefore, regardless of whether the malfunction is in the fat signal or the receptor, the result is obesity. Scientists suspected that the *ob/ob* and *db/db* mice fit into this picture, but they were not sure how.

To determine the causes of the obesity in *ob/ob* and *db/db* mice, investigators connected an *ob/ob* mouse *parabiotically* to a *db/db* mouse; that is, the two mice were physically connected so that the blood of one mouse flowed to the other mouse and vice versa. The *ob/ob* mouse lost weight while the *db/db* mouse remained obese. From these results the investigators concluded that the *ob/ob* mouse lacks the fat signal while the *db/db* mouse lacks the ability to detect the fat signal. The basis for this conclusion is as follows: because the *db/db* mouse is able to produce the fat signal (but cannot respond to it), the fat signal flowed via the circulatory connection to the *ob/ob* mouse where its brain (which can respond to the signal) interpreted the sudden presence of the fat signal as an indication that the days of "starvation" were over. As we now know from experiments performed in the 1990s, the fat signal is a hormone, leptin (from the Greek *leptos,* meaning thin), that is produced by the *OB* gene in fat cells in proportion to the amount of fat they contain.

Leptin is now widely regarded as one of the hormones that play a critical role in keeping us in energy balance. When food is scarce and fat cells are depleted of fat, declining leptin levels signal the brain to increase food intake and decrease metabolism. Conversely, when food is abundant and fat cells are replete with fat, increasing leptin levels signal the brain to decrease food intake and increase metabolism. In other words, as supplies of food wax and wane, behavioral and metabolic adjustments maintain fat levels within a set range.

Despite the symmetry in the body's adjustments to the availability or scarcity of food, among humans obesity is a much more common condition than excessive thinness. This asymmetry in eating and metabolic disorders arises from a combination of factors, including an evolutionary history that placed pressure on our ancestors to eat varied foods that resulted in a balanced diet, but that did not prepare us for the astonishing contemporary increase in the availability of high-fat and highly palatable foods.

Humans are not the only animals that evolved in conditions of scarcity and have trouble adjusting to abundance. Our inability to adjust to the onslaught of high-fat foods is analogous to the reaction of Egyptian spiny mice to freely available water. Spiny mice are desert rodents that possess adaptations that help them to survive in hot, dry environments, including kidneys that are extraordinarily efficient at retaining water. In fact, their kidneys are so efficient that a slice of carrot contains enough moisture to meet their daily water requirements. Veterinarians who oversee the well-being of animals in laboratories must follow state and federal guidelines for their care, and these guidelines rarely take into account the evolutionary histories and physiological adaptations of these animals. Thus these veterinarians typically insist that in laboratories spiny mice, like rats and mice, be given free access to a water bottle. But give free access of water to a spiny mouse and it bloats up like a balloon. Just as evolutionary pressures on humans to deal with the scarcity of food have made us prone to overeating, evolutionary pressures on spiny mice to deal with the scarcity of water have made them prone to overdrinking.

The oversupply of high-fat foods in our modern diet has had disastrous effects for some population groups, groups whose ancestors appear to have coped with alternating periods of feast and famine with adaptations that promote fat storage when times are good. One of the most studied of these groups is the Pima Indians, who have lived in what is now southern Arizona for at least 2,000 years and whose ancestors have lived in North America for 30,000 years. Perhaps as an adaptive response to the uncertainty of food supplies, Pima Indians have very low levels of leptin, a condition that predisposes them to fat storage. This was not a problem a century ago when their diet was high in starch and fiber and only 15 percent of their diet was derived from fat. But times have changed. As the Pima Indians' traditional agrarian lifestyle was disrupted by the diversion of their water supply by American farmers and as assimilation with the larger American society accelerated after World War II, their dietary habits changed dramatically so that today 40 percent of their diet is derived from fat. The result has been a dramatic increase in both obesity and Type II, or late-onset, diabetes.

For Pima Indians as well as for the rest of us, it is easier to gain weight than to lose weight. Nonetheless, most of us maintain reasonably stable body weights throughout our lives regardless of how much we diet or overeat. And even when a diet is successful, success is usually short-lived unless the individual is able to sustain a major change in lifestyle. Contrary to popular opinion, willpower is only part of the problem: there is also a powerful regulatory mechanism at work.

DIT, WAT, AND BAT

We have now seen that when mammals overeat and gain weight, leptin levels rise as fat becomes more abundant. We have also seen that leptin then triggers both a behavioral response, decreased food intake, and a physiological response, increased metabolism. But what exactly do we mean by increased metabolism?

Thirty years ago Michael Stock, working at St. George's Hospital Medical School in London, first noticed that humans who were asked to overeat increased their rates of heat production, a phenomenon called diet-induced thermogenesis, or DIT. Many of us experience the effects of DIT when we feel hot after a large meal; we may even sweat. An increase in heat production after eating could potentially arise for a number of reasons, including the simple fact that digestion is work, and work produces heat as a by-product. But, at least in some cases, DIT seemed to be more than a mere digestive by-product. The heat seemed to be coming from somewhere else. By the end of the 1970s, Stock, in collaboration with Nancy Rothwell, had isolated the source: Brown adipose tissue.

Recall that brown adipose tissue, or BAT, is used by infant mammals and hibernating adults as a means of heat production that replaces or supplements heat production by shivering. In contrast, white adipose tissue, or WAT, consists of fat cells and the energy-rich fat that is contained within them. Stock and Rothwell suggested that BAT is used to burn off excess fat rather than let it move into long-term energy storage as fat. The magnitude of this response is extraordinary: when rats are placed on what is called a *cafeteria diet*—a continuous diet of diverse and highly palatable foods that changes daily—they increase their caloric intake by 75 percent but also increase their metabolism by nearly the same amount, and 90 percent of this increased metabolism is in the form of heat. One of the many implications of this finding is that obesity results from an inability to burn off excess fat, not from mere overeating.

Wasting energy as a means of preventing obesity, however, does not appear to be the whole story. We must also consider the importance of protein, a nutrient that is essential for the normal growth and maintenance of the individual. This is not a problem for many animals, such as the folivorous koala bear, whose only food is the Eucalyptus leaf, or the carnivorous lion, which can eat a zebra without having to take dietary supplements. But the eating habits of omnivores like humans and rats are much more complicated and require that the body keep track of whether we are getting all the kinds of nutrients that we need.

Imagine that a human had access to an abundant supply of low-protein and high-fat foods. If he ate a meal from this source and simply stored the fat from this meal for future energy needs, he would experience increased leptin levels and would therefore stop eating. The result would be a bountiful source of energy in the form of fat but a dangerously low level of protein. But if he burned off the excess fat as heat, he would be driven to eat a second meal from this source, and a third, and a fourth, until he had accumulated the quantity of protein he needed to survive. In other words, DIT appears to be useful not only as a means of avoiding obesity and its attendant health hazards but also for ensuring that we attain balance in our supply of essential nutrients. For humans, when the diet is made up of 12 percent protein, DIT is minimal; for rats the value is 20 percent. Below these values, DIT increases steadily. Above these values, heat production also increases, but as a byproduct of the digestion of high-protein diets; this is not DIT, but it is wasteful. Thus the 12 percent and 20 percent values for the two species represent ideal concentrations of protein for maximally efficient utilization of energy. But such ideal diets are not always available, and many species that typically encounter low-protein diets possess copious amounts of BAT to burn off excess fat and achieve nutrient balance.

We can now return to our *ob/ob* mice and examine the cause of their obesity and the role of leptin in DIT. Consistent with Stock and Rothwell's hypothesis, *ob/ob* mice are obese not because they overeat but because they lack the ability to exhibit DIT. Merely injecting these mice with leptin is sufficient to increase DIT and reduce their obesity. Indeed, leptin's effect on the brain is to activate neurons that stimulate heat production by BAT, in lean and obese animals, thereby burning off fat. Therefore, although widely characterized as a satiety hormone because of its ability to inhibit food intake, leptin can just as reasonably be considered a heat-production hormone because that is, in fact, one of its primary roles—to stimulate BAT thermogenesis.

Investigators have gained additional information about the causes of obesity in rodents by examining its early development. Studies using the *fa/fa* rat (which is analogous to the *db/db* mouse) have provided evidence of deficient heat production by BAT during cold exposure as early as two days of age, before obesity is evident; this deficiency in BAT thermogenesis is even more evident by seven days of age, when pups are already beginning to exhibit signs of obesity. These findings reinforce the view that at least some kinds of obesity reflect a primary malfunction of energy expenditure—and specifically heat production by BAT—because these rat pups are not overfeeding. But perhaps the strongest evidence that the obesity evidenced by the *fa/fa* rat reflects a deficiency in BAT activity comes from a study in which investigators treated *fa/fa* rats, from the age of eight to sixteen days, with a drug that selectively stimulates heat production by BAT. The result of this artificial BAT activation was the prevention of full-blown obesity in these *fa/fa* rats that were otherwise destined to become obese.

The discovery of leptin's ability to promote the burning of excess fat through BAT thermogenesis and to simultaneously suppress eating might seem to indicate that a cure for obesity is near. Indeed, it was hoped that leptin would prove to be an effective treatment for obesity. Unfortunately, most obese humans exhibit increased levels of leptin, suggesting that they are insensitive to leptin and that their predicament is analogous to that of the *db/db* mouse, not that of the *ob/ob* mouse. If so, then leptin supplements are not likely to do much good. However, treatment with leptin may help the 5–10 percent of obese individuals who exhibit low levels of leptin production. And for those of you who are wondering if leptin could help you on your next diet, the answer is "probably," although the side effects of leptin are not yet well understood. On a more practical level, you might have to take out a loan to pay for it—currently, leptin costs nearly $200 per milligram.

LEAN TIMES

Although the average American has been getting fatter and fatter for the last few decades, the American ideals of beauty have increasingly shifted toward thinness. Whereas a healthy woman should carry approximately 25 percent of her body weight as fat, most of the women we see in movies or in magazines carry only 10–15 percent of their weight as fat. This discrepancy between the physical realities faced by most women and the messages portrayed by a minority of women who are so thin that many of them no longer have menstrual cycles has helped to generate a steady increase in the incidence of anorexia nervosa over the last twenty years.

First described in 1689 by an English physician, Robert Morton, as "nervous consumption," anorexia nervosa was given its official name in a paper published in 1873 by another English physician, William Gull. Today anorexia nervosa afflicts approximately 1 percent of all adolescent and young adult females in industrialized countries. Its victims are ten times more likely to be women than men; these women are typically from developed countries, between fifteen and twenty-five years of age, and from middle- or upper-class families. The fourth edition of the Diagnostic and Statistical Manual of Mental Disorders, the bible of psychopathology, defines anorexia according to a number of criteria, including refusal to maintain body weight within 85 percent of the expected body weight given an individual's age and height, an intense fear of becoming fat, and denial of the seriousness of the problem. These are just threshold diagnostic criteria. Some anorexics lose up to 60 percent of their body weight.

There have been many attempts to identify the cause of anorexia, most of them focusing on the psychological features of the disorder. For example, it has been noted that many people who develop anorexia are perfectionists, introverts, and insecure. Perhaps the most widely known explanation is that anorexics have a distorted "body image" or "body concept," viewing themselves as overweight when they are actually emaciated. Research has indicated, however, that overestimation of body size is not a likely explanation of the cause of the disorder.

Sometimes the best way to identify the cause of a behavior or disorder is to simply observe its development in real time. And it is important that the observer not be the patient, whose insights are often highly questionable. My wife occasionally tells people that she quit smoking one day and that was it. Such stories are common. But what the tellers typically leave out of these stories are descriptions of their numerous attempts and failures, of the process of teaching themselves to ignore their bodily cravings and to train their bodies to live for longer and longer periods of time without the externally provided nicotine. Is it possible that anorexics have kicked the eating habit in the same way that ex-smokers have quit the smoking habit?

Dieting is, by its very nature, a process of learning to ignore internal bodily signals that it is time to eat. The ability to learn to diet is not something that evolution would have favored. Nonetheless, with our relatively large brains and extraordinary capacity to learn, and confronted with the modern abundance of high-fat foods, many people have learned to restrain their eating, and some have taken this to the extreme of starving themselves. Indeed, the notion that anorexia is the unwanted by-product of dieting has a firm foundation. First, eating disorders like anorexia are most common among the groups of people who do the most dieting. Second, up to 20 percent of anorexics were previously obese, a condition that is likely to lead to dieting. Finally, anorexia is common in professions, such as modeling, ballet dancing, and wrestling, in which dieting plays a prominent role. Whether these relations between dieting and anorexia are causal or correlational has not yet been determined.

Regardless of the causes of anorexia nervosa, it is quite clear that its victims have starved themselves into an emaciated state. Their refusal to eat and to appreciate the seriousness of their predicament and the stated desire of some that they would rather die than eat suggest that neither mind nor body is functioning appropriately. One might even be tempted to conclude that the problem is entirely in the mind—if they would only eat, their problem would go away. Such a conclusion, however, would be about as helpful as telling an obese person whose brain lacks the ability to detect leptin that he should just stop eating.

Although anorexia nervosa is typically considered a psychological disorder and is therefore treated as such, it is also known that anorexics have a variety of physiological abnormalities that may contribute to the onset of the disorder, the maintenance of the disorder, or both. One such abnormality concerns a number of dimensions of thermoregulatory functioning. A large proportion of anorexics (40–80 percent) have body temperatures below normal and complain frequently of feeling cold, suggesting that they are hypothermic. The starvation-induced reduction in subcutaneous fat certainly decreases their ability to retain heat, but they also show reduced metabolism as well as tendencies to sweat and to dilate skin blood vessels at lower-than-normal temperatures, suggesting that they are regulating at a lower-than-normal body temperature. When experimentally exposed to hot or cold temperatures, anorexics fail to activate appropriate thermoregulatory responses, and their body temperatures increase or decrease passively, as would a reptile's under similar circumstances. And when they do eat, anorexics complain of excessive heat within their bodies as well as in their hands and feet, and they even sweat.

Despite these profound disturbances of thermoregulation, there is little reliable information about the nature of the thermoregulatory deficit in anorexia. Attempts by investigators to explain the thermoregulatory deficits in anorexia are largely unconvincing: there simply are not enough solid, reliable data. We do know that leptin levels and basal metabolism are very low in anorexics, which is consistent with their state of starvation but does not explain why these patients do not respond to the reduced leptin signal by increasing food intake. Perhaps, as some have suggested, anorexia is caused by a physiological abnormality induced by a brain malfunction.

Our understanding of the thermoregulatory state of anorexics was further confused by the recent finding that severely malnourished anorexics exhibit exaggerated increases in diet-induced thermogenesis when they eat a meal. In fact, as much as 50 percent of the energy content of meals eaten by anorexics is "wasted" as heat, and this finding helps to explain the difficulties that anorexics have in regaining body weight despite their reduced metabolisms. But this finding of exaggerated DIT in anorexics also suggests that, from a physiological standpoint, anorexics are regulating their amount of body fat at a dangerously low level just as most of the rest of us are regulating our amount of body fat at a normal level.

It is important to emphasize that there may be a distinct difference between the factors that drive anorexics to a state of semi-starvation and the factors that prevent them from recovering. The former issue is, at present, unresolved. It is possible to make some sense of the available data, but some reinterpretation will be necessary. On our trip from here to there, we will first need to make yet another stop in ratland.

In general, like homeowners turning down their thermostats during an energy crisis, animals reduce their heat production during times of starvation. For example, when infant rats are starved (by being separated from their mother for more than ten hours) and subsequently exposed to cold, they fail to activate heat production by BAT and body temperature falls passively (this is similar to the response of anorexics to decreasing air temperature). This failure to activate heat production is not, as one might think, merely a reflection of dwindling energy supplies. As it turns out, during starvation, but before all stored energy has been utilized, infant rats actively suppress heat production by BAT and allow their body temperature to fall in the cold. This suppression is controlled by the brain, whose role is to assess the energetic state of the animal and respond accordingly.

It had been known for some time that a starved rat pup needs only a single meal to regain its ability to produce heat in the cold. My colleagues and I wondered, however, whether leptin plays a role in regulating the infant rat brain's response to starvation and cold. To test this idea, we separated pups less than one week of age from their mothers for eighteen hours in a warm incubator, after which we gave some of them a dose of leptin and tested their thermoregulatory capabilities. The pups given leptin did not exhibit fully normal responses, which is not surprising given that they had been starved for eighteen hours, but they did exhibit significantly more heat production by BAT than did pups that were starved but not given leptin. What this experiment showed is that a hypothermic animal given access to leptin can be tricked into increasing its heat production.

The connection to anorexia is straightforward. Anorexics live in a state of semi-starvation and, like infant rats, are protecting their meager energy reserves by reducing metabolism and body temperature. As already stated, they commonly complain of feeling cold, even when housed in a room that most people would consider thermally comfortable. Therefore, when anorexics eat a meal, their exaggerated increase in heat production is probably due to two factors: diet-induced thermogenesis and increased heat production to raise their body temperature back to normal (just as a starved infant rat given leptin increases heat production in the cold). This would mean that anorexics are not merely dumping their food energy as waste heat (as in diet-induced thermogenesis) but are taking advantage of an infusion of energy to warm up.

Regardless of whether an anorexic exhibits diet-induced thermogenesis or whether she is simply trying to warm up, from the standpoint of the patient's health the primary desire of converting food energy into fat so that the patient can gain weight has been thwarted. There is, however, a simple way to manipulate the amount of energy that is wasted: first, provide the patient with the choice of several thermal environments as well as other opportunities to behaviorally thermoregulate (for example, electric blanket, gloves, heat lamp, hot tub); second, when she is thermally comfortable, provide her with a variety of foods and allow her to continue to adjust to her thermal needs. Both forms of heat production involved here—diet-induced thermogenesis and heat production in the cold—will be influenced by the thermal state of the patient. If she is warm before eating, then less of the food energy will be wasted and more will be converted to fat. Indeed, William Gull, working with anorexics over 125 years ago, emphasized the need to supply external heat to his patients when they ate. Such treatments fell into disfavor during the twentieth century with the uninhibited growth of the psychopathology industry, but there are encouraging signs on the horizon: the

Swedish neuroendocrinologists Cecilia Bergh and Per Södersten are reporting impressive success rates with anorexics using a treatment that, in the spirit of Gull, involves control of the patient's thermal environment. In addition, the development of an animal model of anorexia nervosa—called activity-based anorexia—holds additional hope for new insights into the nature of the disorder in humans.

It bears repeating that there are many possible causes of anorexia nervosa, from psychopathology, to dieting gone awry, to hypothalamic dysfunction. Indeed, even if psychological explanations have merit, they may have relatively little bearing on the treatment of anorexics once they have fallen into a state of semi-starvation. The thermal factors described here represent only a part of the anorexic's physiological difficulties. But even so, anorexics are suffering through profound metabolic and thermal disturbances that may very well be the cause of their difficulty in recovering from their condition. Research on these disturbances is still rare, but fortunately this situation appears to be changing. And if, as Bergh and Södersten contend, psychopathology is an effect, not a cause, of the starvation that characterizes anorexia, increased attention to the physiological causes and consequences of this terrifying disorder can only help.

9 THE LIGHT GOES OUT

An elderly man has just settled down for a twenty-minute nap. He is seated upright in a cushioned chair, a small metal ball cradled in the upturned palm of his hand. The hand extends beyond the arm of the chair, and the ball is poised to fall into a metal kettle on the floor below. His wrist and forearm are rigid to support the ball. In fact, all of the muscles in his body are actively working to maintain postural support. As the man settles into his nap, his neck muscles start to relax and his head tilts to the side. His breathing slows. His wrist and arm begin to relax. Suddenly, quietude is replaced with activity—erratic breathing and darting eyes beneath the closed lids. But this activity is quickly interrupted. His wrist and fingers have lost their strength, and the ball rolls out of his hand and tumbles into the kettle with a loud clanking sound.

Thomas Alva Edison awakes refreshed, lifts himself out of the chair, and continues with his day's work. Edison had discovered a natural and easy method to limit his naptime—a discovery that even this great inventor did not fully understand.

On average, we spend one-third of our lives asleep. Most of us sleep in one large chunk of approximately eight hours, but Edison eschewed such wasteful nonsense; he promoted his meager sleep regimen as a virtue that reflected his commitment to work. "Sleep is like a drug," he once said. "Take too much at a time and it makes you dopey." In fact, we all have different needs for sleep that remain quite stable throughout our lifetimes. These needs are highly idiosyncratic and apparently have little bearing on our intelligence or productivity. Albert Einstein, for example, slept ten hours each night. Enough said.

Edison's nap-limiting technique appears to have been motivated by a desire to save time, but he probably did not realize that he was also limiting the character of the sleep he was getting. Specifically, Edison was selectively preventing himself from entering a form of sleep known as rapid eye movement (REM) sleep and thereby restricting his naps to a form of sleep known as non-REM (NREM) sleep. Because NREM sleep typically precedes REM sleep (whether during a nap or at night), and because one of the characteristics of the transition from

NREM to REM sleep is complete relaxation of all the muscles of the body, the clatter of the falling metal ball provided auditory feedback to Edison that he had made the transition to REM sleep. But he did not understand the significance of this transition, because REM sleep would not be discovered by the scientific world until 1953, twenty-two years after Edison's death.

REM sleep gets its name from one of its many features, namely the rapid twitching movements of the eyes. It was the discovery of these movements that resulted in the eventual realization that there were two distinct forms of sleep. Since then, hundreds of investigators have devoted their careers to understanding the function, neural control, pharmacology, development, and evolution of REM and NREM sleep. We have learned much, but not nearly enough to satisfy even the most complacent sleep researcher. More important, there is still no consensus among researchers as to why we sleep. Indeed, we still don't really know what sleep is.

The numerous mysteries of sleep have been the subject of many books, but our aim here is more modest. We seek only to understand why temperature plays such a predominant role in whether or not we get a good night's sleep. Why do we sleep so shallowly and awake so unrefreshed when our bedroom is too hot or too cold? And why does our sleepiness seem to wax and wane like the tides rather than increase continuously over time? As we will see, the answers to these questions have relevance for a number of real-life experiences, from camping out in the cold to jet lag.

SNUG AS A BUG IN A RUG

In his work *On Sleep and Sleeplessness (De somno et vigilia),* Aristotle (384–322 BCE) noted that "fits of drowsiness are especially apt to come on after meals." On the basis of this and other observations, he suggested that sleep and waking are influenced by the movement of heat and cold throughout the body as a result of digestion. His intuition that temperature is involved in sleep was a good one; he just happened to get the details a little wrong. So let's address the details.

We know that all animals seek a thermally comfortable environment before lying down to sleep. When such an environment is unavailable, animals will use physiological and behavioral means to achieve thermal balance. On cold days, dogs and cats will curl up into tight balls, drawing in their legs and tail so as to prevent heat loss. On warm days, they will sprawl out, perhaps on a cool tile floor if one is available, attempting to maximize heat loss to the environment. These thermoregulatory postures are maintained and adjusted to the prevailing air temperature during NREM sleep. Therefore, sleeping posture during NREM sleep is molded by the local thermal environment.

Humans are a bit more picky than dogs and cats about where and when we sleep. We have rooms and furniture dedicated to this activity. We place blankets (maybe even electric blankets) and down comforters on our beds to keep us warm. If we don't have air conditioning, we may place a fan near our bed on a hot night. And as we get older and more attuned to our physical needs, we learn a few tricks that help us to compensate for our own thermal shortcomings. Maybe we realize that we prefer a slightly colder bedroom temperature, or perhaps we learn to wear socks or slippers to prevent cold feet. Over the course of a lifetime, our sleeping habits become rituals that we must follow if we are, like Edison, to awake refreshed and ready to go.

Our thermoregulatory behavior does not stop when we get into bed and fall asleep. As we cycle through NREM and REM sleep, a cycle that humans repeat approximately six times in a normal night, we awake briefly (although we are usually not aware of these awakenings), adjust our posture, and kick off or pull up the covers. The sleep period, in other words, is not a time of completely quiet relaxation.

In fact, REM sleep is not a time of relaxation at all. In addition to rapid eye movements, small muscle twitches in the limbs, irregular breathing, and irregular heartbeat, REM sleep is characterized by electrical activity in the cerebral cortex (the outer layer of the brain) that resembles the activity observed in an awake person. In part, this activity gives rise to the bizarre dreams that accompany REM sleep, but REM sleep is much more than simply a period of or for dreaming. It is a period of intense activity throughout many parts of the body and brain. That this activity occurs in a sleeping person whose muscles are relaxed struck many early investigators as paradoxical, and, indeed, an alternative designation of REM sleep is *paradoxical sleep.*

Despite all the activity associated with REM sleep, there is also plenty of inactivity. As already mentioned, the muscles are relaxed. And muscle tone is not all that relaxes. During REM sleep we and other mammals also inhibit our thermoregulatory responses to cold and warm stimulation, in effect placing ourselves at the mercy of the prevailing thermal environment. For example, when a cat settles down for a nap in an environment that is warm enough to produce panting but not hot enough to prevent slumber, the panting will continue even as the cat descends into NREM sleep. If REM sleep is entered, however, panting ceases, resulting in decreased heat loss and increased body temperature. In humans in a hot environment, sweating is reduced during REM sleep, also resulting in increased body temperature.

Thermoregulatory responses to the cold are also inhibited during REM sleep. A cold cat will shiver to maintain body temperature when it is awake, and this shivering will continue during NREM sleep, but shivering will stop if the cat enters REM sleep, resulting in increased heat loss and decreased body temperature. In addition, adult rats inhibit heat production by brown adipose tissue during REM sleep.

Experimenters can mimic an animal's thermoregulatory responses to heating and cooling by manipulating the temperature of its hypothalamus. But even manipulation of hypothalamic temperature is incapable of overcoming the inhibition imposed by REM sleep. For example, although selective heating of the hypothalamus is sufficient to trigger panting in a cat while it is awake or in NREM sleep, such heating is ineffective once the cat is in REM sleep. Thus, through a mechanism that is not yet understood, the REM sleep state has the effect of disengaging the hypothalamus and other brain regions involved in thermoregulation and thereby disrupting the normal control of body temperature.

Because of the inhibition of thermoregulatory responses in animals during REM sleep, body temperature will rise or fall according to the prevailing thermal environment. These fluctuations will be more pronounced in small animals, with their relatively large surface area, than in larger animals. But regardless of body size it appears that mammals only tend to make the transition from NREM sleep to REM sleep within a set range of air temperatures. It is as though a gatekeeper is protecting us from the thermal vulnerability imposed by REM sleep.

When the air temperature poses little danger, the gate is lifted and REM sleep is entered. When the temperature is too hot or too cold, however, the gate is not lifted. Instead, we toss and turn or curl up in a ball, sweat or shiver vigorously, throw off or pull up the covers—anything to get comfortable. If we fail to achieve an acceptable body temperature, we may doze off at times and gain some NREM sleep, but we will wake up in the morning unrefreshed and groggy because of our inability to enter REM sleep and benefit from its restorative effects.

We don't know why thermoregulation is so dramatically inhibited during REM sleep. But we don't know a lot of things about REM sleep, such as why our breathing and heart rates become irregular or why our eyes dart beneath our closed lids or why our cerebral cortex becomes active. There are many theories, some of them silly and some of them intriguing. At the moment, what we can conclude with certainty is that temperature is one of the few major factors that limit both the quantity and the quality of sleep in adult mammals.

Although the vast majority of research on sleep is conducted on adults, it is infants who sleep the most. Moreover, REM sleep is the predominant form of sleep in many infants, especially in altricial infants like kittens and puppies. For example, kittens spend 40 percent of each day in REM sleep and 5 percent in NREM sleep, while the corresponding numbers for cats are 16 percent REM and 52 percent NREM. The predominance of REM sleep during early development is also seen in humans. In human infants born ten weeks prematurely, REM sleep occupies 80 percent of the infant's total sleep time; this value drops to 60–65 percent in infants born two weeks prematurely and to 50 percent in full-term infants (who sleep sixteen hours each day). This trend continues until REM sleep occupies 25 percent of total sleep time by ten years of age, after which it stabilizes and changes little before declining further in the elderly.

This striking life-history pattern of REM sleep has inspired a number of theories regarding the role of sleep, and especially REM sleep, in the development of the central nervous system and its neural connections. Regardless of which theory may turn out to be most accurate, it is quite clear that infants require a stable thermal environment to grow and mature; hence the ubiquity of incubators in maternity wards and neonatal intensive care units. The importance of a warm thermal environment is twofold: first, a cold infant must expend energy staying warm, thereby diverting this energy from the infant's primary occupations, growth and maturation; second, even more than in adults, who have larger bodies and are more accomplished thermoregulators, infant sleep is sensitive to air temperature. For example, in infant hamsters the amount of

REM sleep decreases in lockstep with the temperature of the air. Therefore, to get enough REM sleep, infants are critically dependent on a comfortable thermal environment.

WE'VE GOT RHYTHMS

When he invented a practical method for producing light using electricity, Thomas Edison changed our lives forever. The effortless extension of the day beyond the limits previously established by the Sun began a process that has seriously disrupted the behavioral rhythms that once defined our lives. Anyone who has experienced the crushing sleepiness that accompanies jet lag or the severe stress that develops after too many graveyard shifts knows that these rhythms are not mere habit. We dance to these rhythms whether we want to or not.

Perhaps the most dominant and stable of all bodily rhythms is body temperature. The value we commonly give for human body temperature, 37°C (98.6°F), is, like a broken clock, accurate only twice a day. The rest of the day, our body temperature cycles between approximately 36° and 38°C (97° and 100°F), reaching its lowest point in the early morning around the time when we typically wake up and its highest point in the late evening when we typically go to sleep. We know that this temperature rhythm is truly internally driven—and not merely a by-product of rest and activity—because it is exhibited by people who are bedridden. Moreover, the temperature rhythm is driven primarily by changes in rates of heat loss through the skin rather than by changes in heat production.

The relationship between our pattern of sleeping and waking and our body temperature raises the obvious question of whether the two rhythms are causally related. Our best insights into this question have come from studies in which human subjects have agreed to live alone in underground bunkers or isolated apartments for months at a time, completely disconnected from the outside world. At first the experimenters may control the apartment's lights, turning them off at 11 P.M. and on at 7 A.M. With this externally provided cue, or zeitgeber (from the German, meaning time-giver), the subject's sleep and temperature rhythms remain in synchrony. But when the experiment is changed so that the subject is allowed to turn the lights in the room on and off at will, the subject's link to the outside world is completely severed, leaving nothing but biology to sort things out.

So what happens to the clock underlying the body's temperature rhythm? The rhythm is still very regular, but the clock starts to run a little fast, gaining anywhere from twelve minutes to over an hour each day (there is some disagreement on this point). Therefore, the role of the zeitgeber—whether it be light or any other external time cue—is to reset the clock each day, just as we would turn back a clock that runs a little fast. In the absence of a zeitgeber, our internal clock is now referred to as a *free-running clock.* The next question is obvious: What happens to the sleep rhythm when the internal clock is free-running? Does it remain locked to the temperature rhythm, or does it move to the beat of a different drummer?

The answer is a little of both. When human subjects begin to free-run, two basic patterns are seen. The majority of subjects remain internally synchronized to the body temperature rhythm, but now they begin their sleep periods near the temperature minimum and wake up approximately eight hours later on the rising portion of the temperature rhythm (recall that normal subjects typically begin sleeping at the temperature maximum in the evening and wake up at the temperature minimum in the morning).

Approximately one-third of the free-running subjects, however, exhibit a mixed response, referred to as internal desynchronization. On some days (remember, these are days defined by the subjects' own internal rhythms, not by a twenty-four-hour clock), a subject may go to sleep at the temperature minimum and wake up eight hours later as body temperature is increasing. But on other days the same subject may go to sleep near the body temperature maximum (as most of us normally do), sleep through the temperature minimum, and wake up some fourteen hours later, again at the rising portion of the temperature rhythm. In other words, in relation to the body temperature rhythm, onset of sleep varies much more than time of waking, resulting in highly variable durations of sleep. Moreover, subjects who experience these highly variable sleep patterns cannot distinguish between a one-hour nap and a fourteen-hour crash.

Body temperature is only one indicator of the rhythm of the internal clock. For example, the hormone melatonin is released into the bloodstream primarily during the night in most mammals, including humans, and it appears to have some sleep-promoting effects. Administration of melatonin reduces body temperature, and some believe that this thermal effect underlies melatonin's promotion of sleep. The scientific basis for the casual use of melatonin is still in dispute. Unfortunately, this has not prevented melatonin from becoming the only hormone that can be obtained in the United States without a prescription. It is now being regularly consumed by millions of Americans at ridiculously high doses for a variety of sleep complaints. Only time will tell what unfortunate consequences may result from this fad.

The rhythms exhibited by body temperature, melatonin release, and other hormonal and physiological systems can be traced to a small part of the hypothalamus called the *suprachiasmatic nucleus,* or SCN. Thanks to work conducted over the last two decades, we now know that the SCN acts as an internal clock, keeping time for the body even in the absence of external cues. In fact, when neurons from the SCN are removed from the brain and kept alive in a dish, they continue to exhibit time-keeping properties. Furthermore, there are direct neural connections between the eyes and the SCN, just as one would expect given the importance of light as a zeitgeber in humans and other animals.

The identification of the SCN as one of the internal clocks that drive many of our biological rhythms, including body temperature, helps to explain why we become so discombobulated after international flights that carry us eastward or westward, but not after flights going northward or southward. When we travel across time zones, our internal clock needs time to readjust to the external cues in the new environment, and meanwhile there is a tension between our internal clock and the rhythms of life that prevail at our destination. The same tension exists for shift workers, who must continually struggle with a schedule of work and sleep that is out of sync with the signals from their biological clocks. There is even evidence that insomnia (a problem that afflicts approximately 8 percent of the adult population and a quarter of the elderly population at any given time) and depression are associated with shifts and other alterations in body temperature rhythms. Whether these changes in body temperature contribute to causing these problems or are merely symptoms of a more general malfunction in SCN rhythmicity is not yet known.

As Winston Churchill once said about the Soviet Union, "It is a riddle wrapped in a mystery inside an enigma." The same could be said of sleep. Although we have learned much about sleep through the work of hundreds of researchers over many decades, some of the most basic aspects of sleep's development, neural control, and function continue to elude us. Nonetheless, we do know that temperature and thermoregulation play important roles in modulating a variety of sleep-related variables, including the onset, duration, and quality of sleep. Some investigators have been so impressed by the interrelatedness of sleep and temperature that they have proposed that sleep serves thermoregulatory functions. It has not yet been determined whether such hypotheses have merit. But, as we have seen throughout this book, temperature is interrelated with many important aspects of animal physiology and behavior, and we can only benefit from continued research on its role in molding and modifying our lives of sleep and sleeplessness.

On October 19, 1931, the world read of the death of Thomas Edison. According to a United Press obituary published in *The Washington Post,* the man whose invention of the incandescent light bulb forever altered the pace and rhythm of our lives "passed peacefully from a deep sleep into death." The obituary notes that the great inventor, who worked such long hours and who disdained sleep, lay in serene repose at his home in West Orange, New Jersey, with nothing but "the occasional purr of an automobile to disturb the silence in which they permitted him to rest." The headline above this dramatic obituary was, appropriately enough, "The Light Goes Out." But despite its aptness for Edison, the *Post* headline mischaracterizes the nature of life. Although visible light, which represents but a sliver of the spectrum of electromagnetic radiation, undoubtedly beautifies our world, it is the heat absorbed from solar and non-solar radiation that makes life happen. Even in the Pacific Ocean, at depths where not a single photon from the Sun can penetrate, hot thermal vents have become oases for organic life. In other words, life is more about heat than about light.

EPILOGUE

JOHN UPDIKE once composed, for a Brazilian newspaper, an essay entitled "The Cold." Brazilians, it seems, were in need of some insight into the nature of cold, and who better to provide this insight than a man, gifted with language, who has experienced life year-round in New England? In his essay, Updike notes that "cold is an absence, an absence of heat, and yet it feels like a presence." This simple statement beautifully captures the asymmetry of the laws of thermodynamics—that heat flows naturally only down, not up, a thermal gradient—as well as the sense that these laws fail to capture the harsh experiential reality of the cold.

Whether cold is a presence or an absence, humans have worked strenuously to protect themselves from its consequences. Over time, our cultures have incorporated and magnified this fundamental biological need into something much more. Although shelters were initially built as mere barriers against heat loss, they eventually became endothermic, producing heat as well. Home and hearth. Houses with fireplaces are still more highly valued, even if that enhanced value has more to do with the ambiance fireplaces create than with the warmth they provide. Perhaps someday we will gather with our families around a flat-screen panel that projects the visual and auditory sensations of a fire. But even such an odd cultural development will maintain the thread that connects us to those times when our ancestors were continually threatened by the harshness of the outside world.

As we saw in Chapter 2, the thermal gradients on the surface of our planet have structured the migratory patterns of its many inhabitants. Temperature also structures our day-to-day activities. Updike says that, in Europe, "the statistics for readership go down as the latitude becomes southerly; a warm climate invites citizens outdoors, to the sidewalk café, the promenade, the brain-lulling beach." The first great universities of Europe—in Bologna, Italy, and Salamanca, Spain—were therefore particularly impressive achievements for the gathering of scholars and students in hot climates. During the Middle Ages, as centers of learning flourished in the colder climates of northern Europe, scholars cloistered themselves in dank and damp surroundings, buffered from the cold by robes, hoods, and hats that to this day are recalled symbolically by the caps and gowns of graduation ceremonies.

This book's epigraph is a quotation from the final paragraph of Updike's essay. He ends this paragraph with a note of optimism and implied recognition of the ineluctable interconnection between warmth and life: "To return back indoors after exposure to the bitter, inimical, implacable cold is to experience gratitude for the shelters of civilization, for the islands of warmth that life creates." "Islands of warmth that life creates": this phrase captures the ability of humans and other animals to provide for their thermal needs by manipulating their local environment to meet those needs: shelters, be they houses or nests, are behavioral arrangements that enhance our chances for survival when our world turns against us. But such behavioral capabilities are recent arrivals, for the availability of warmth on our planet far predates our ability to control and manipulate it. Thus Updike's phrase is a half-truth: Earth, in fact, is an island of life that warmth has created.

BIBLIOGRAPHY, CREDITS

ACKNOWLEDGMENTS

INDEX

BIBLIOGRAPHY

For each chapter, references are provided to scientific reports that were used as source material for the text. For general readers interested in delving more deeply into a particular subject, books that I have found particularly accessible and enjoyable are marked with asterisks.

1. Temperature: A User's Guide

*Atkins, P. W. *The second law.* New York: Scientific American Books, 1984.

Calder, W. A., III. *Size, function, and life history.* Cambridge, Mass.: Harvard University Press, 1984.

*Haldane, J. B. S. *On being the right size and other essays.* Oxford: Oxford University Press, 1985.

Korb, J., and K. E. Linsenmair. The effects of temperature on the architecture and distribution of *Macrotermes bellicosus* (Isoptera, Macrotermitinae) mounds in different habitats of a West African Guinea savanna. *Insectes Sociaux,* 45 (1998): 51–65.

Lüscher, M. Air-conditioned termite nests. *Scientific American,* 205 (1961): 138–145.

Schmidt-Nielsen, K. *Desert animals: Physiological problems of heat and water.* Oxford: Oxford University Press, 1964.

*Schmidt-Nielsen, K. *Scaling: Why is animal size so important?* Cambridge: Cambridge University Press, 1984.

Schmidt-Nielsen, K., B. Schmidt-Nielsen, S. A. Jarnum, and R. T. Houpt. Body temperature of the camel and its relation to water economy. *American Journal of Physiology,* 188 (1957): 103–112.

*Shachtman, T. *Absolute zero and the conquest of cold.* New York: Houghton Mifflin, 1999.

*von Baeyer, H. C. *Warmth disperses and time passes.* New York: Modern Library, 1998.

2. Behave Yourself

Alberts, J. R. Huddling by rat pups: Group behavioral mechanisms of temperature regulation and energy. *Journal of Comparative and Physiological Psychology,* 92 (1978): 231–245.

Blumberg, M. S., S. J. Lewis, and G. Sokoloff. Incubation temperature modulates post-hatching thermoregulatory behavior in the Madagascar ground gecko, *Paroedura pictus.* Submitted for publication.

*Diamond, J. *Guns, germs, and steel.* New York: Norton, 1999.

Evans, R. M. Vocal regulation of temperature by avian embryos: A laboratory study with pipped eggs of the American white pelican. *Animal Behaviour,* 40 (1990): 969–979.

Fraenkel, G. S., and D. L. Gunn. *The orientation of animals.* 1940; New York: Dover, 1961.

Hawkins, B. A. Ecology's oldest pattern? *Trends in Ecology and Evolution,* 16 (2001): 470.

Heinrich, B. Energetics of honeybee swarm thermoregulation. *Science,* 212 (1981): 565–566.

Korycansky, D. G., G. Laughlin, and F. C. Adams. Astronomical engineering: A strategy for modifying planetary orbits. *Astrophysics and Space Science,* 275 (2001): 349–366.

Le Maho, Y., P. Delclitte, and J. Chatonnet. Thermoregulation in fasting emperor penguins under natural conditions. *American Journal of Physiology,* 231 (1976): 913–922.

Mori, I., and Y. Ohshima. Neural regulation of thermotaxis in *Caenorhabditis elegans. Nature,* 376 (1995): 344–348.

Satinoff, E. Behavioral thermoregulation in the cold. In *Handbook of Physiology,* ed. M. J. Fregly and C. M. Blatteis, pp. 481–505. Oxford: Oxford University Press, 1996.

Wilkins, L., and C. P. Richter. A great craving for salt by a child with cortico-adrenal insufficiency. *Journal of the American Medical Association,* 114 (1940): 866–868.

Yadin, Yigael. *Masada: Herod's fortress and the zealots' last stand.* New York: Random House, 1966.

Yegül, F. *Baths and bathing in classical antiquity.* New York: Architectural History Foundation, 1992.

3. Then Bake at 98.6°F for 400,000 Minutes

Carvalho, M., F. Carvalho, and M. L. Bastos. Is hyperthermia the triggering factor for hepatotoxicity induced by 3,4-methylenedioxymethamphetamine (ecstasy)? An in vitro study using isolated mouse hepatocytes. *Archives of Toxicology,* 74 (2001): 789–793.

Crews, D. Animal sexuality. *Scientific American,* 270 (Jan. 1994): 108–114.

Deeming, D. C., and M. W. J. Ferguson. *Egg incubation: Its effects on embryonic development in birds and reptiles.* Cambridge: Cambridge University Press, 1991.

Germain, M.-A., W. S. Webster, and M. J. Edwards. Hyperthermia as a teratogen: Parameters determining hyperthermia-induced head defects in the rat. *Teratology,* 31 (1985): 265–272.

Malberg, J. E., and L. S. Seiden. Small changes in ambient temperature cause large changes in 3,4-methylenedioxymethamphetamine (MDMA)-induced serotonin neurotoxicity and core body temperature in the rat. *Journal of Neuroscience,* 18 (1998): 5086–94.

Nijhout, H. F. Metaphors and the role of genes in development. *BioEssays,* 12 (1990): 441–446.

Oyama, S., P. E. Griffiths, and R. D. Gray, eds. *Cycle of contingency: Developmental systems and evolution.* Cambridge, Mass.: MIT Press, 2001.

4. Everything in Its Place

Baker, M. A. Influence of the carotid rete on brain temperature in cats exposed to hot environments. *Journal of Physiology,* 220 (1972): 711–728.

Baker, M. A. A brain-cooling system in mammals. *Scientific American,* 240 (1979): 130–139.

Baptiste, K. E., J. M. Naylor, J. Bailey, E. M. Barber, K. Post, and J. Thornhill. A function for guttural pouches in the horse. *Nature,* 403 (2000): 382–383.

Bligh, J., and K. Voigt, eds. *Thermoreception and temperature regulation.* Berlin: Springer-Verlag, 1990.

Block, B. A. Structure of the brain and eye heater tissue in marlins, sailfish, and spearfishes. *Journal of Morphology,* 190 (1996): 169–189.

Block, B. A., et al. Migratory movements, depth preferences, and thermal biology of Atlantic bluefin tuna. *Science,* 293 (2001): 1310–14.

Cabanac, M., and M. Caputa. Natural selective cooling of the human brain: Evidence of its occurrence and magnitude. *Journal of Physiology,* 286 (1979): 255–264.

Heinrich, B. A. *The hot-blooded insects: Strategies and mechanisms of thermoregulation.* Cambridge, Mass.: Harvard University Press, 1993.

Kilgore, D. L., M. H. Bernstein, and D. M. Hudson. Brain temperature in birds. *Journal of Comparative Physiology* B, 110 (1976): 209–215.

Pinshow, B., M. H. Bernstein, G. E. Lopez, and S. Kleinhaus. Regulation of brain temperature in pigeons: Effects of corneal convection. *American Journal of Physiology,* 242 (1982): R577–R581.

Satinoff, E. Behavioral thermoregulation in response to local cooling of the rat brain. *American Journal of Physiology,* 206 (1964): 1389–94.

Schmidt-Nielsen, K. Countercurrent systems in animals. *Scientific American,* 244 (1981): 118–128.

Schmidt-Nielsen, K. *Animal physiology: Adaptation and environment.* Cambridge: Cambridge University Press, 1990.

Seymour, R. S. Plants that warm themselves. *Scientific American,* 276 (1997): 104–109.

Seymour, R. S., and P. Schultze-Motel. Thermoregulating lotus flowers. *Nature,* 383 (1996): 305.

Waites, G. M. H. The effect of heating the scrotum of the ram on respiration and body temperature. *Quarterly Review of Biology,* 47 (1962): 314–323.

5. Cold New World

Alberts, J. R., and G. J. Decsy. Terms of endearment. *Developmental Psychobiology,* 23 (1990): 569–584.

Bass, M., R. E. Kravath, and L. Glass. Death-scene investigation in sudden infant death. *New England Journal of Medicine,* 315 (1986): 100–105.

Blix, A. S., and J. W. Lentfer. Modes of thermal protection in polar bear cubs—at birth and on emergence from the den. *American Journal of Physiology,* 236 (1979): R67–R74.

Blumberg, M. S. The developmental context of thermal homeostasis. In E. M. Blass, ed., *The Handbook of Behavioral Neurobiology,* vol. 13: *Developmental Psychobiology, Developmental Neurobiology and Behavioral Ecology: Mechanisms and Early Principles,* 199–227. New York: Plenum, 2001.

Blumberg, M. S., and G. Sokoloff. Do infant rats cry? *Psychological Review,* 108 (2001): 83–95.

Gunn, T. R., K. T. Ball, G. G. Power, and P. D. Gluckman. Factors influencing the initiation of nonshivering thermogenesis. *American Journal of Obstetrics and Gynecology,* 164 (1991): 210–217.

Hochachka, P. W. Defense strategies against hypoxia and hypothermia. *Science,* 231 (1986): 234–241.

Lagercrantz, H., and T. A. Slotkin. The "stress" of being born. *Scientific American,* 254 (1986): 100–107.

Mendelsohn, E. *Heat and life: The development of the theory of animal heat.* Cambridge, Mass.: Harvard University Press, 1964.

Schwab, S., S. Schwarz, M. Spranger, E. Keller, M. Bertram, and W. Hacke. Moderate hypothermia in the treatment of patients with severe middle cerebral artery infarction. *Stroke,* 29 (1998): 2461–66.

Seidler, F. J., and T. A. Slotkin. Adrenomedullary function in the neonatal rat: Responses to acute hypoxia. *Journal of Physiology,* 358 (1985): 1–16.

Smith, R. E., and B. A. Horwitz. Brown fat and thermogenesis. *Physiological Reviews,* 49 (1969): 330–425.

Stanton, A. N. Overheating and cot death. *Lancet* (1984): 1199–1201.

Willinger, M. SIDS: A challenge. *Journal of NIH Research,* 1 (1989): 73–80.

6. Fever All through the Night

Bernheim, H. A., and M. J. Kluger. Fever and antipyresis in the lizard, *Dipsosauraus dorsalis. American Journal of Physiology,* 231 (1976): 198–203.

Cooper, K. E. *Fever and antipyresis.* Cambridge: Cambridge University Press, 1995.

Houk, J. C. Control strategies in physiological systems. *FASEB Journal,* 2 (1988): 97–107.

Kluger, M. J. *Fever: Its biology, evolution, and function.* Princeton: Princeton University Press, 1979.

Kluger, M. J. Is fever beneficial? *Yale Journal of Biology and Medicine,* 59 (1986): 89–95.

Kluger, M. J., D. H. Ringler, and M. R. Anver. Fever and survival. *Science,* 188 (1975): 166–168.

Mrosovsky, N. *Rheostasis: The physiology of change.* Oxford: Oxford University Press, 1990.

Stitt, J. T. Fever versus hyperthermia. *FASEB Journal,* 38 (1979): 39–43.

7. The Heat of Passion

Anderson, C. A. Heat and violence. *Current Directions in Psychological Science,* 10 (2001): 33–38.

Blumberg, M. S., J. A. Mennella, and H. Moltz. Hypothalamic temperature and deep body temperature during copulation in the male rat. *Physiology and Behavior,* 39 (1987): 367–370.

Blumberg, M. S., and H. Moltz. How the nose cools the brain during copulation in the male rat. *Physiology and Behavior,* 43 (1988): 173–176.

Caterina, M. J., A. Leffler, A. B. Malmberg, W. J. Martin, J. Trafton, K. R. Petersen-Zeitz, M. Koltzenburg, A. I. Basbaum, and D. Julius. Impaired nociception and pain sensation in mice lacking the capsaicin receptor. *Science,* 288 (2000): 306–313.

Caterina, M. J., M. A. Schumacker, M. Tominaga, T. A. Rosen, J. D. Levine, and D. Julius. The capsaicin receptor: A heat-activated ion channel in the pain pathway. *Nature,* 389 (1999): 783–784.

Dib, B. Effects of intracerebroventricular capsaicin on thermoregulatory behavior in the rat. *Pharmacology, Biochemistry, and Behavior,* 16 (1982): 23–27.

McClintock, M. K. Group mating in the domestic rat as a context for sexual selection: Consequences for the analysis of sexual behavior and neuroendocrine responses. *Advances in the Study of Behavior,* 14 (1984): 1–50.

Montesquieu. *The spirit of laws.* 1748; Berkeley: University of California Press, 1977.

8. Livin' off the Fat

American Psychiatric Association. *Diagnostic and statistical manual of mental disorders.* 4th ed. Washington, 1994.

Bergh, C., and P. Södersten. Anorexia nervosa: Rediscovery of a disorder. *Lancet,* 351 (1998): 1427–29.

Blumberg, M. S., K. Deaver, and R. F. Kirby. Leptin disinhibits nonshivering thermogenesis in infants after maternal separation. *American Journal of Physiology,* 276 (1999): R606–R610.

Brownell, K. D., and C. G. Fairburn, eds. *Eating disorders and obesity: A comprehensive handbook.* New York: Guilford, 1985.

Charon, C., F. Dupuy, and R. Bazin. Effect of the ß-adrenoceptor agonist BRL-35135 on development of obesity in suckling Zucker *(fa/fa)* rats. *American Journal of Physiology,* 268 (1995): E1039–E1045.

Friedman, J. M., and J. L. Halaas. 1998. Leptin and the regulation of body weight in mammals. *Nature,* 395: 763–770.

Gull, W. Anorexia nervosa (apepsia hysterica, anorexia hysterica). *Transactions of the Clinical Society of London,* 7 (1874): 22–28.

Krahn, D. D., C. Rock, R. E. Dechert, K. K. Nairn, and S. A. Hasse. Changes in resting energy expenditure and body composition in anorexia nervosa patients during refeeding. *Journal of the American Dietetic Association,* 93 (1993): 434–438.

Moore, B. J., S. J. Armbruster, B. A. Horwitz, and J. S. Stern. Energy expenditure is reduced in preobese 2-day Zucker *fa/fa* rats. *American Journal of Physiology,* 249 (1985): R262–R265.

Moukaddem, M., A. Boulier, M. Apfelbaum, and D. Rigaud. Increase in diet-induced thermogenesis in severely malnourished anorexia nervosa patients. *American Journal of Clinical Nutrition,* 66 (1997): 133–140.

Planche, E., M. Joliff, P. De Gasquet, and X. Leliepvre. Evidence of a defect in energy expenditure in 7-day-old Zucker rat *(fa/fa). American Journal of Physiology,* 245 (1983): E107–E113.

Stock, M. J. Gluttony and thermogenesis revisited. *International Journal of Obesity,* 23 (1999): 1105–17.

Woods, S. C., R. J. Seeley, D. Porte Jr., and M. W. Schwartz. Signals that regulate food intake and energy homeostasis. *Science,* 280 (1998): 1378–83.

9. The Light Goes Out

Blumberg, M. S., and D. E. Lucas. A developmental and component analysis of active sleep. *Developmental Psychobiology,* 29 (1996): 1–22.

Czeisler, C. A., et al. Stability, precision, and near-24-hour period of the human circadian pacemaker. *Science,* 284 (1999): 2177–81.

Glotzbach, S. F., and H. C. Heller. Temperature regulation. In *Principles and practice of sleep medicine,* ed. M. H. Kryger, T. Roth, and W. C. Dement, pp. 289–304. Philadelphia: W. B. Saunders, 2000.

Jouvet-Mounier, D., L. Astic, and D. Lacote. Ontogenesis of the states of sleep in rat, cat, and guinea pig during the first postnatal month. *Developmental Psychobiology,* 2 (1970): 216–239.

Sokoloff, G., and M. S. Blumberg. Active sleep in cold-exposed infant Norway rats and Syrian golden hamsters: The role of brown adi-

pose tissue thermogenesis. *Behavioral Neuroscience,* 112 (1998): 695–706.
*Winfree, A. T. *The timing of biological clocks.* New York: Scientific American Library, 1986.
Zully, J., R. Wever, and J. Aschoff. The dependence of onset and duration of sleep on the circadian rhythm of rectal temperature. *Pflügers Archive,* 391 (1981): 314–318.

Epilogue

Updike, John. The cold. In *More matter: Essays and criticism,* pp. 133–135. New York: Knopf, 1999.

CREDITS

Figure 1. Heat exchange. Adapted from V. G. Dethier et al., *Topics in the study of life: The BIO source book* (Harper and Row, 1971). Used by permission of Pearson Education, Inc.

Figure 3. Masada. Adapted from Yigael Yadin, *Masada: Herod's fortress and the Zealots' last stand,* copyright © 1966 by Yigael Yadin. Used by permission of Random House, Inc.

Figure 4. Movements of nematode worms. Adapted from I. Mori and Y. Ohshima, "Neural regulation of thermotaxis in *Caenorhabditis elegans,*" *Nature,* 376 (1995): 344–348. Used by permission.

Figure 5. Emperor penguin chick. Photograph by Franz Lanting/Minden Pictures.

Figure 6. The sacred lotus. Photograph by Minhtien Tran. Used by permission.

Figure 7. Polar bear mother and cub. © 2000 T. Davis/W. Bilenduke/Stone.

Figure 8. Obese mouse and normal mouse. Courtesy of The Jackson Laboratory, Bar Harbor, Maine.

ACKNOWLEDGMENTS

My research interests in temperature began in graduate school at the University of Chicago, so it is appropriate to acknowledge the intellectual foundations that were laid there. Jeff Alberts provided me with the perfect postdoctoral environment at Indiana University, and I can never thank him sufficiently for his guidance, support, and continued friendship.

I owe an enormous debt of gratitude to Richard Panek and the other writers in his Nonfiction Narrative workshop at the Iowa Summer Writing Festival for encouraging me to finish the book. Leaving the comfortable world of academia to focus on writing with writers, even for just a week, was wonderful, and it was my fellow students, and especially Richard, who gave me the confidence to carry through to the end.

Some of the research discussed in this book was done in collaboration with colleagues and students, and I especially want to acknowledge the contributions of Greta Sokoloff and Robert Kirby. Thanks are also due my colleagues in the

Department of Psychology at the University of Iowa for their friendship, support, and advice. Each year the International Society for Developmental Psychobiology provides a forum for scientists and their students to present their work, and my research has benefited enormously from the stimulation and inspiration that that society fosters.

Because research is an expensive enterprise, I appreciate the generous support for my work from the National Institute of Mental Health and the National Institute of Child Health and Human Development. A Faculty Scholar Award from the University of Iowa provided the freedom to complete the book.

The thoughtful comments of Jo McCarty, Margie Blumberg, Herschel Blumberg, and Greta Sokoloff substantially improved the book, as did the comments of two anonymous reviewers. I am indebted to Elizabeth Knoll of Harvard University Press for giving the book a chance and for shepherding me so patiently and wisely through what was a very unfamiliar process. Camille Smith, also of Harvard University Press, is an editor that many authors only dream about.

My family has been a continual source of support in all possible ways. To my parents, Goldene and Herschel, my sisters, Susan and Margie, my nephews, Aaron and Ari, and especially my wife, Jo—my deepest thanks for your love and understanding.

INDEX